T0140581

Learning and Analytics in Intelligent Systems

Volume 17

The main aim of the series is to make available a publication of books in hard copy form and soft copy form on all aspects of learning, analytics and advanced intelligent systems and related technologies. The mentioned disciplines are strongly related and complement one another significantly. Thus, the series encourages cross-fertilization highlighting research and knowledge of common interest. The series allows a unified/integrated approach to themes and topics in these scientific disciplines which will result in significant cross-fertilization and research dissemination. To maximize dissemination of research results and knowledge in these disciplines, the series publishes edited books, monographs, handbooks, textbooks and conference proceedings.

More information about this series at http://www.springer.com/series/16172

Pradeep Kumar Mallick · Prasant Kumar Pattnaik ·
Amiya Ranjan Panda · Valentina Emilia Balas
Editors

Cognitive Computing in Human Cognition

Perspectives and Applications

 Springer

Editors
Pradeep Kumar Mallick
School of Computer Engineering
Kalinga Institute of Industrial Technology
(KIIT) Deemed to be University
Bhubaneswar, Odisha, India

Prasant Kumar Pattnaik
School of Computer Engineering
Kalinga Institute of Industrial Technology
(KIIT) Deemed to be University
Bhubaneswar, Odisha, India

Amiya Ranjan Panda
School of Computer Engineering
Kalinga Institute of Industrial Technology
(KIIT) Deemed to be University
Bhubaneswar, Odisha, India

Valentina Emilia Balas
Department of Automatics and Applied
Software, Faculty of Engineering
Aurel Vlaicu University of Arad
Arad, Romania

ISSN 2662-3447   ISSN 2662-3455   (electronic)
Learning and Analytics in Intelligent Systems
ISBN 978-3-030-48120-9   ISBN 978-3-030-48118-6   (eBook)
https://doi.org/10.1007/978-3-030-48118-6

This Springer imprint is published by the registered company Springer Nature Switzerland AG
The registered company address is: Gewerbestrasse 11, 6330 Cham, Switzerland

Preface

The aim of the book is to bring together leading academic scientists, researchers, and research scholars to exchange and share their experiences and research results on all aspects of cognitive computing in human cognition. It also provides a premier interdisciplinary platform for researchers, practitioners, and educators to present and discuss the most recent innovations, trends, and concerns as well as practical challenges encountered and solutions adopted in the fields of IoT and analytics for agriculture. The book is organized into ten chapters,

Chapter "Improved Steganography Using Odd Even Substitution": The proposed image steganography technique implements an LSB technique with an algorithm to hide message bits in the DCT coefficients.

Chapter "A Tags Mining Approach for Automatic Image Annotation Using Neighbor Images Tree": This chapter developed a new model for addressing these issues using a tree mechanism of photos related to the target image.

Chapter "A Survey: Implemented Architectures of 3D Convolutional Neural Networks": In this paper, we survey on different implementations of 3D convolutional neural networks and their respective accuracies for different datasets. We then compare all the architectures to find which one is most suitable to perform flexibly on CBCT-scanned images.

Chapter "An Approach for Detection of Dust on Solar Panels Using CNN from RGB Dust Image to Predict Power Loss": This chapter focuses on CNN-based approach to detect dust on solar panel and predicted the power loss due to dust accumulation. We have taken RGB image of solar panel from our experimental setup and predicted power loss due to dust accumulation on solar panel.

Chapter "A Novel Method of Data Partitioning Using Genetic Algorithm Work Load Driven Approach Utilizing Machine Learning": The proposed chapter utilizes natural computing optimization-inspired genetic algorithm (GA) for the improvisation of the partitioned data structure. The optimized set is cross-validated utilizing artificial neural network.

Chapter "Virtual Dermoscopy Using Deep Learning Approach": This chapter presents an automated dermatological diagnostic system using a deep learning approach. Dermatology is the branch of medicine which deals with the identification and treatment of skin diseases. The presented system is a machine interference in contradiction to the traditional medical personnel-based belief of dermatological diagnosis.

Chapter "Evaluating Robustness for Intensity Based Image Registration Measures Using Mutual Information and Normalized Mutual Information": This chapter uses information measures such as mutual information (MI) and normalized mutual information (NMI) to obtain the aligned image and then evaluated their robustness.

Chapter "A New Contrast Based Degraded Document Image Binarization": This chapter proposes the binarization technique which uses the contrast feature to compute the threshold value with minimum parameter tuning. It computes the local contrast image using maximum and minimum pixel values in the neighborhood. The high contrast text pixels in the image are detected using global binarization.

Chapter "Graph Based Approach for Image Data Retrieval in Medical Application": This chapter describes new technique such as max flow graph-based approach. Max flow-based techniques have more accurate result than PCA and Hessian method.

Chapter "Brain Computer Interface: A New Pathway to Human Brain": This chapter focuses on past 15 years, and this assistive technology has attracted potentials numbers of users as well as researchers from multidiscipline.

We are sincerely thankful to Almighty to supporting and standing at all times with us, starting from the call for chapters till the finalization of chapters, all the editors have given their contributions amicably, which it is a positive sign of significant team works. The editors are sincerely thankful to all the members of Springer. We are equally thankful to reviewers for their timely help and authors who have shared their chapters.

Bhubaneswar, India   Pradeep Kumar Mallick
Bhubaneswar, India   Prasant Kumar Pattnaik
Bhubaneswar, India   Amiya Ranjan Panda
Arad, Romania   Valentina Emilia Balas

About This Book

This edited book designed the cognitive computing in human cognition to analyze to improve the efficiency of decision making by cognitive intelligent. The book also intended to attract the audience who work in brain computing, deep learning, transportation, and solar cell energy. Due to this in recent era, smart methods with human touch called as human cognition are adopted by many researchers in the field of information technology with the cognitive computing.

Key Features

1. Addresses the complete functional framework workflow in advances of cognitive computing.
2. Addresses the different data mining techniques was applied.
3. Exploring data studies related to data-intensive technologies in solar panel energy, machine learning, and big data.

Contents

About the Editors

Dr. Pradeep Kumar Mallick is currently working as a senior associate professor in the School of Computer Engineering, Kalinga Institute of Industrial Technology (KIIT) Deemed to be University, Odisha, India. He has also served as a professor and head of the Department of Computer Science and Engineering, Vignana Bharathi Institute of Technology, Hyderabad. He has completed his postdoctoral fellow (PDF) in Kongju National University, South Korea, Ph.D. from Siksha 'O' Anusandhan University, M. Tech. (CSE) from Biju Patnaik University of Technology (BPUT), and MCA from Fakir Mohan University Balasore, India. Besides academics, he is also involved in various administrative activities, Member of Board of Studies, Member of Doctoral Research Evaluation Committee, Admission Committee, etc. His area of research includes algorithm design and analysis, and data mining, image processing, soft computing, and machine learning. Now he is the editorial member of Korean Convergence Society for SMB. He has published nine books and more than 70 research papers in national and international journals and conference proceedings in his credit.

Dr. Prasant Kumar Pattnaik, Ph.D. (Computer Science), Fellow IETE, Senior Member IEEE is a professor at the School of Computer Engineering, KIIT Deemed University, Bhubaneswar. He has more than a decade of teaching and research experience. He has published numbers of research papers in peer-reviewed international journals and conferences. He also published many edited book volumes in Springer and IGI Global Publication. His areas of interest include mobile computing, cloud computing, cyber security, intelligent systems, and brain–computer interface. He is one of the associate editors of *Journal of Intelligent & Fuzzy Systems*, IOS Press and Intelligent Systems Book Series Editor of CRC Press, Taylor Francis Group.

Dr. Amiya Ranjan Panda has seven years of research experience in DRDO and one year of teaching experience. He received the B.Tech. degree in information technology from Biju Patnaik University of Technology, Rourkela, India, in 2009, and the M.Tech. degree in computer science and engineering from the Kalinga

Institute of Industrial Technology (KIIT), Bhubaneshwar, India, in 2012. He received Ph.D. degree from Siksha 'O' Anusandhan University, Bhubaneshwar, in the year 2017 working in a real-time project "Design, Development and Implementation of Software Defined Radio-based Flight Termination System" of DRDO, ITR, Chandipur.

Dr. Valentina Emilia Balas is currently a full professor in the Department of Automatics and Applied Software at the Faculty of Engineering, "Aurel Vlaicu" University of Arad, Romania. She holds a Ph.D. in Applied Electronics and Telecommunications from Polytechnic University of Timisoara. She is the author of more than 280 research papers in refereed journals and international conferences. Her research interests include intelligent systems, fuzzy control, soft computing, smart sensors, information fusion, modeling, and simulation. She is the editor-in-chief of the *International Journal of Advanced Intelligence Paradigms* and the *International Journal of Computational Systems Engineering*. She also serves as an editorial board member of several national and international journals and is the director of Intelligent Systems Research Center in Aurel Vlaicu University of Arad. She is a member of EUSFLAT, SIAM, and a Senior Member IEEE, member in TC—Fuzzy Systems (IEEE CIS), member in TC—Emergent Technologies (IEEE CIS), and member in TC—Soft Computing (IEEE SMCS).

Improved Steganography Using Odd Even Substitution

Ramandeep Kaur Brar and Ankit Sharma

Abstract Image steganography is one of the emerging techniques used for hiding content in digital images. This makes the content more secure and free from information attackers. The proposed image steganography technique implements an LSB technique with an algorithm to hide message bits in the DCT coefficients. The technique has chosen a mid-band frequency area for the bit's substitution in an odd and even bit selection. The results are better than traditional LSB. The parameters used are PSNR and MSE to understand the blur or noise added to the cover image while adding the secret message.

Keywords QR code · Steganography · Image steganography · Least significant bit (LSB) technique · Discrete cosine transform (DCT) · Peak signal to noise ratio (PSNR) · Mean square error (MSE)

1 Introduction

1.1 QR Code

QR code located for Quick Response Code that is a kind of matrix barcode otherwise we can say that two-dimensional barcode which is intended for the first manufacturing in Japan. Mainly, a barcode is a machine-readable photosensitive tag that comprises data around the piece to which it is involved [1]. A QR code consists of black modules such as square dots arranged in a square on a white backdrop set, which can be deciphered by an imaging device like a camera and the process of with error correction rate when the image does not go well, explained. In both horizontal and

R. K. Brar (✉) · A. Sharma
Computer Science and Engineering, Baba Farid College of Engineering and Technology, Bathinda, India
e-mail: brar.raman111@gmail.com

A. Sharma
e-mail: ankit.bfcet@gmail.com

© Springer Nature Switzerland AG 2020
P. K. Mallick et al. (eds.), *Cognitive Computing in Human Cognition*,
Learning and Analytics in Intelligent Systems 17,
https://doi.org/10.1007/978-3-030-48118-6_1

Fig. 1 Module structure [1]

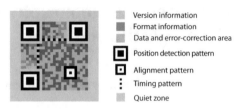

vertical sections of the image, the required data are then separated from the given pattern [1, 2].

1.2 Code Structure

The QR Code segments run many jobs: Approximately genuine facts keeps itself, whereas others are divided into different job outlines that recover interpretation presentation and permit mark configuration, fault improvement and misrepresentation recompense. The timing patterns know the amount of sign to the scanning device. Near is also an essential "quiet zone," a four-module inclusive bumper zone in which there is no information, to confirm that nearby text or symbols are not wrong for QR Code data [2] (Fig. 1).

Through scanning hardware, search patterns of the location can be searched and can increase the overall speed by reading image and data processing [1].

1.3 Steganography

Steganography is a method of implanting confidential communications in such a system that anyone else can see the message without knowing it. Along with the bits of various classified messages, replacing the bits of fewer frequently data used in data records like graphics, audio, image, etc. implies Steganography. This secret information can also be straightforward text, ciphertexts or images [3] (Fig. 2).

Fig. 2 Overview of steganography techniques [4]

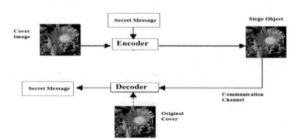

1.4 Least Significant Bit (LSB) Technique

In LSB technology, the first image is divided into 3 planes i.e. red, green and blue. Each plane has the pixel representation of the image that has a fixed intensity values (Fig. 3).

LSB substitution replaces the binary bit from the secret message with the message bits from secret message image bit sequence. The value at LSB place is either '0' or '1'. Occasionally the new bit after replacing is the same as that of older value. In this case, replacing can be takes place but the alteration in the image parameters is not reflected. This adds to privileged PSNR value in stego image [7, 6].

2 Literature Survey

Vladimír Hajduk et al. [4], This paper is focuses on schematic planning of image steganographic method which can embed encrypted secret message using Quick Response Code (QR) code into image data. For embedding the QR code, the use of the Crete Wavelet Transformation (DWT) domain is used while the embedding process is furthermore confined by the Advanced Encryption Standard (AES) cipher algorithm. Moreover, usually features of QR code was busted using the encryption, thus it makes the system further protected. The effectiveness of the planned method was measured by Peak Signal-to-Noise Ratio (PSNR) and accomplished results were equated with other steganographic tackles.

Kumawat et al. [1], In this paper, author works in the latest automatic technology concepts. In the current past, the concept of Quick Response Code (QR Code) has attained a considerable recognition and is being used as data representation mean. This paper providing complete information on all the concepts of Quick Response Code. The first experimentation is to be relevant noise in QR Code (encoding) and second is De-noising (decoding) using Median and Wiener filters. This document also affords a glance of the impact of Noise on the QR code and Histogram of the PSNR values that shows the evaluation of the images.

Chaturvedi et al. [8], This framework presents a new proposal for whacking a logo-based watermark in color still image. This move is based on averaging of middle frequency coefficients of block Discrete Cosine Transform (DCT) coefficients of an image. Here we suggest an algorithm of aliquot watermarking technique based on

Fig. 3 Least significant bit insertion (LSB) [5, 6]

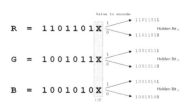

DCT (Discrete Cosine Transformation) using mid band for toughness. We use the DCT mid band to use different image formats with co-efficiency and the analysis seemed that the JPG image formats PSNR value on an average was lowest in different with others.

Mehboob et al. [9], Due to its simplicity, the most commonly used system is the use of least significant bits. The least significant bit or its format is usually used to hide data in a digital image. This paper discusses the science of Steganography in all-purpose and proposes a work of fiction procedure to hide data in a colorful image using the least significant bit.

3 Methodology

The transform domain-based technique "DCT" helps to implant the hidden secret message effectively as we know that transform domain technique is far further better than the spatial domain technique in the meantime toughness in contradiction of lossy compression and different filtering options such as median, high-pass and low-pass filters etc. [10].

This procedure gives a better declaration about the superiority on the receiving ends. The mid band frequency quantities of an 8 into 8 DCT block. The DCT Technique divides the image into blocks of 8 into 8 [10]. Low frequency (FL) is used to indicate the lower frequency coefficients of the block whereas high frequency (FH) is used to indicate the higher frequency coefficients. By avoiding significant modifications of the cover image the mid band is worn for embedding to provide additional resistance to lossy compression techniques [11] (Fig. 4).

The methodology explains the comprehensive working of the image steganography process. In the projected technique, as implanting a secret message image into shipper image [12], two files are required. First file is the colored image, known as cover image and the next file brings the message itself to be concealed [12] (Fig. 5).

The technique hides data in QR code cover image (Fig. 6).

To get a high PSNR, the self-hidden message has been changed to 0- and 1-bit order [14]. This bit order is divided into two groups that are odd group and even group. The splitting of the message bit progression has improved spatial distribution between the pixels of the message inside the cover image which in spin reduces the noise level in the cover-image [15, 16].

Fig. 4 DCT region for
Mid-frequency [11]

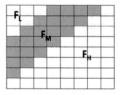

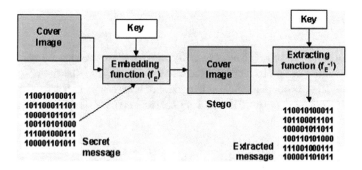

Fig. 5 Steganography overview [13]

Fig. 6 Bit sequence broken to even and odd groups [13]

3.1 Algorithm

Phase-I Algorithm used to place in data image into the cover-image:

Step 1.   To load the cover image.
Step 2.   To load the Message Image.
Step 3.   To convert text message/image into binary bit sequence and divide into two groups of even and odd.
Step 4.   To calculate and locate LSB of each one pixel of the cover image.
Step 5.   To replace each bit of message image with the Mid Band of the cover image.
Step 6.   To save the steganographed image.

Phase-II Algorithm used to preserve data image from the cover image:

Step 1.   is to load the Stegnographed image.
Step 2.   is to calculate and find the LSB of each pixel of the cover image containing the message image.
Step 3.   is to recover message bit stream from the cover image.
Step 4.   is to determine peak signal to noise ratio of Cover Image.

3.2 *Parameters*

Peak Signal to Noise Ratio (PSNR): Usually, the image steganography system should insert the content of an obscured message in the image such that the graphic excellence of the image is not perceptibly changed. For that reason, to learn the embedding perceptual effect, we have used the peak signal to noise ratio (PSNR) which is defined as [5, 16]:

$$\text{PSNR} = 10 \log_{10} \frac{(L-1)^2}{\text{RMS}}$$

where

$$\text{RMS} = \frac{1}{m \cdot n} \sum_{i=1}^{m} \sum_{j=1}^{n} \left(x_{i,j} - x'_{i,j} \right)^2$$

Mean Square Error (MSE): In a sense, the resolution of a spreading center should be linked to an error. If we say that the number t is a best evaluate of center, then most likely we are proverb that t represents the whole spreading better, in some way than other numbers [17]. In this context, assume that we measure the superiority of t, as a quantity of the center of the distribution, in terms of the mean square error

$$\text{MSE}(t) = \frac{1}{n} \sum_{i=1}^{k} f_i (x_i - t)^2 = \sum_{i=1}^{k} p_i (x_i - t)^2$$

MSE(t) is a partial average of the squares of the distances between t and the class marks with the relative frequencies as the weight factors. Therefore, the best evaluates of the center, comparative to this measure of error, is the value of t that minimizes MSE. In statistics, the mean square error (MSE) is one way to analyze the variance amongst an estimator and the accurate value of the amount being predictable [13].

4 Results

Result 1
    See Table 1.
Result 2
    See Table 2.
Result 3
    See Table 3.

Table 1 Results of image steganography in QR code

S. No.	Payload	Payload size	Cover image	Message image	PSNR	MSE
1	Wow wowwowo wow owowowowowo	31B		PayT M	42.4477	0.0582816

Table 2 Results of image steganography in QR code

S. No.	Payload	Payload size	Cover image	Message image	PSNR	MSE
2	;uioehjkhjkshadkhlasjkdhaslwkejrkwjelwkqje	42B			42.4302	0.0585162

Table 3 Results of image steganography in QR Code

S. No.	Payload	Payload size	Cover image	Message image	PSNR	MSE
3	Pin number 7892463578036578 dataapasstor	41B		Hi	42.4937	0.0576669

References

1. D. Kumawat, R.K. Singh, D. Gupta, S. Gupta, Impact of denoising using various filters on QR code. *International Journal of Computer Applications* **63**(5) (2013)
2. M.W. Islam, A novel QR code guided image stenographic technique. IEEE International Conference on Consumer Electronics (ICCE). IEEE (2013)
3. B.C. Nguyen, S.M. Yoon, H.-K. Lee, *Multi-bit plane image steganography*. Department of EECS, Korea Advanced Institute of Science and Technology, Guseong-dong, Yuseong-gu, Daejeon, Republic of Korea (2006)
4. V. Hajduk, O. Martin Broda, L. Dušan, Image stegano-graphy with using QR code and cryptography, in *26th International Conference, Radioelektronika (RADIOELEKTRONIKA)*. IEEE (2016), pp. 350–353
5. A. Basit, M. Y. Javed, *Iris localization via intensity gradient and recognition through bit planes*. Department of Computer Engineering, College of Electrical and Mechanical Engineering, National University of Sciences and Technology (Nust), Peshawar Road, Rawalpindi, Pakistan
6. AMS Rahma, M.E. Abdulmunim, R.J.S. Al-Janabi, New spatial domain steganography method based on similarity technique. Int. J. Eng. Technol. **5**(1) (2015)
7. B.G. Banik, S.K. Bandyopadhyay, Review on steganography in digital media. *International Journal of Science and Research (IJSR)* 4.2 (2015)
8. R. Chaturvedi, A. Sharma, N. Hemrajani, D. Goyal, Analysis of robust watermarking technique using mid band DCT domain for different image formats. International Journal of Scientific and Research Publications **2**(3), 1–4 (2012)
9. B. Mehboob, R.A. Faruqi, A steganography implementation. In *International Symposium on Biometrics and Security Technologies, 2008*, ISBAST 2008. IEEE (2008), pp. 1–5
10. B.H. Barhate, R.J. Ramteke, Comparative analysis and measuring qualitative factors using colour and gray scale images for LSB and DCT-mid band coefficient digital watermarking. Int. J. Comput. Appl. **128**(12), 34–38 (2015)

11. H. Sheisi, J. Mesgarian, M. Rahmani, Steganography: DCT coefficient replacement method and compare with JStegAlgorithm. Int. J. Comput. Electr. Eng. **4**(4) (2012)
12. A.S. Pandit, S.R. Khope, Faculty Student, Review on image steganography. Int. J. Eng. Sci. 6115 (2016)
13. G. Manikandan, R. Jeya, *A steganographic technique involving JPEG bit-stream.* Department of Computer Science and Engineering SRM University, Kattan-Kulathur, Kancheepuram, Tamil Nadu, India. IJEDR, 3(2) (2015)
14. A. Mondal, S. Pujari, A novel approach of image-based steganography using pseudorandom sequence generator function and DCT coefficients. Int. J. Comput. Netw. Inf. Secur. (2015)
15. V. Saravanan, A. Neeraja, Security issues in computer networks and steganography, in *2013 7th International Conference on Intelligent Systems and Control (ISCO)*. IEEE (2013)
16. Anshit Agarwal, Stretching the limits of image steganography. Int. J. Sci. Eng. Res. **5**(2), 1253–1256 (2014)
17. A.K. Gulve, M.S. Joshi, A high capacity secured image steganography method with five pixel pair differencing and LSB substitution. Int J Image Graph Signal Process (2015)
18. Aditi Soni, Sujit K. Badodia, Implementation of improved steganography for hiding text on digital data. Int. J. Sci. Res. (IJSR) **4**(11), 80–84 (2015)
19. F. Luthon, M. Lievin, F. Faux, On the use of entropy power for threshold selection. Signal Process. (2004)

A Tags Mining Approach for Automatic Image Annotation Using Neighbor Images Tree

Vafa Maihami

Abstract With the growth of the internet and the volume of images, both online (such as Flicker and Facebook) and offline (such as image datasets or personal/organizational collections), in recent years, annotation of the image has taken broad attention. Image annotation, is a method where labels or keywords for an image are created. This may be biased to popular labels in the automatic image annotation relying on the closet neighbors. Furthermore, when confronting images with less common and unique keywords, the efficiency of these methods will reduce. This article developed a new model to addressing these issues using a tree mechanism of photos related to the target image. Firstly, focused on the correlation of the feature vector between the target image and the image database, neighbor's images are obtained in the form of a matrix. Afterwards, to indicate the different tags of interest, the non-redundant tags subspaces are mined. Next, based on the tags it has in common with the query image, a neighbor's images tree structure is drawn. Finally, recommendation tags to the query image are obtained through using the tree structure. The Proposed method applied to well-known benchmarks of annotated datasets, Corel5k, IAPR TC12 and MIR Flickr. The results of the experiments indicate that the method suggested is the better performance and improves the results in comparison with the other methods.

Keywords Relevance labels · Content based image retrieval · Structure tree · Similarity measure · Automatic image annotation

1 Introduction

This enormous increase in use of social networks and visual items has led to urgent development demands and interpretation of multimedia content. Thus, it is necessary to extend new solutions to analyze these data in novel programs. During the last years, the image annotation method has been considered as one of the strong method to

V. Maihami (✉)
Department of Computer Engineering, Sanandaj Branch, Islamic Azad University, Sanandaj, Iran
e-mail: Maihami@iausdj.ac.ir

© Springer Nature Switzerland AG 2020   9
P. K. Mallick et al. (eds.), *Cognitive Computing in Human Cognition*,
Learning and Analytics in Intelligent Systems 17,
https://doi.org/10.1007/978-3-030-48118-6_2

manage the visual contents [1, 2]. Image annotation is method to evaluate the visual contents of an image and apply the tags to it [1]. Image annotation may be applicable in different fields such as industry, medicine, art, etc. [3–8].

Many tools for achieving of automatic image annotation methods such as learning machines and interaction between people and computers have been offered [9–14]. However, some of these approaches are limited in accuracy, complexity and scalability. The number of approaches is based on segmentation techniques which normally cannot break down images into the respective regions [15, 16]. Similar labels/tags are commonly used to identify similar images. Accordingly, the nearest-neighbor-based image annotation has recently attracted attention [17–22]. While image annotation rely on the nearest-neighbor does not achieve explicit generalization, test images are compared to training instances. Since the assumptions are built directly from the training examples themselves, this technique is called the nearest neighbor.

In past works on image annotation relying on the nearest neighbors, common nonparametric methods have been used where voting is used to assign tags from visually similar items [17, 20]. For multimodal content and vast vocabulary of weak labels, the property of this technique is accepted but it is very vulnerable to the measure used to identify visual neighbors. New studies have employed learnable methods and weighted voting structures [19, 23, 24], or the images used for voting were more carefully selected [25]. Although this approach presents good results, they are limited because they treat tags and visual features in separate ways. Often, they may be biased to popular labels. In addition, they drop in results if they confronting to images with less similar and unique labels.

In this paper, the above mentioned challenges have been addressed by introducing a new structure tree of the neighbor's images. Firstly, labels of adjacent images are obtained in the form of a matrix based on the similarity of the feature vector between the input image and the dataset instances. Afterwards, the subspaces of the non-redundant tags are recognized and the system is in an attempt to find a list of neighbors which are sharing common tags on each subspace. The neighbors belonging to the same subspace have the same interest tags. Obviously, a unique neighbor image may relate to a few lists with various labels. The lists can be used to construct a novel structure tree of the neighbor's images. The target image is usually placed at the tree root, and the other neighbor images are put at the tree's other locates. The place of each neighbor is defined by its labels and the similarity that the label has with the target image. The proposed structure resulted in a neighbor image with no tag in common with the query image can get a place in the tree. Lastly, the recommended tags to the query image are obtained by using the tree structure.

Our findings can be defined as follows: (1) an efficient structure for solving bias to popular labels is presented, which their efficiency is reduces when they confronted with less similar and unique labels in the image annotation based on nearest neighbors. (2) The non-redundant subspaces are mined to display the tags of the image. (3) Unique labels are extracted with the query image and a tree structure is generated. (4) In addition, to showing the success of our process, experimental results on famous datasets are presented for evaluating the efficiency of the suggested method.

The remaining of the article is organized as follows: in Sect. 2, the related works are presented. Next, the developed framework is outlined in Sect. 3. Section 4 describes experimental results of the presented approach, and the conclusion is set out in Sect. 5.

2 Related Works

Although image annotation approaches may be done both manually and automatically, automatic image annotation is actually more focused. Manually image tagging is time consuming, costly and relies on the person's vision. Automatic annotation is an idea of using computers in place of humans. Precision is worse than manually but it is much quicker with lower cost. Various techniques have been presented for automatic image annotation [19, 20, 26–29]. In past years, major research resources have also been dedicated to image annotation [21, 30, 31]. Typically, automatic image annotation methods can be divided into the two class [1, 30, 32]:

- Search-based methods: In these methods, the labels are provided explicitly by using images in the database. The k-nearest neighbor (KNN) method (such as the related techniques) with varying distances is generally used due to convenience and good efficiency in large scale datasets.
- Learning-based methods: These methods are challenging because they are known a multi-class/binary classification. The limitation of these methods is that modeling the associations among labels is not easy, and will face problems. after applying to a high number of labels.

The sparse representation technique and its variants have drawn the notice of scientists and have demonstrated successful for plenty of vision purposes, especially automatic image annotation [13, 29, 33]. In recent years, efficient learning algorithms based on deep learning techniques and neural networks has presented and been applied to different fields like computer vision, signal recognition, etc. [28, 32]. In [34] challenges of training deep architectures are reviewed and in paper [35] the investigations on various strategies to create a better recent approaches of deep convolutional neural network image categorization are carried out. The main problems of these approaches are time and space complexities, in addition, there are no theoretical support with deep learning.

A new survey shows that tags gathered from neighbor's images are as useful as tags provided by some Learning-based models for explaining image content [21]. As mentioned in the introduction, simple nonparametric models were used in previous works on image annotation based on the nearest neighbor in which voting was used to exchange labels among Similar in appearance photos [17, 20]. For multimodal content and vast vocabulary of weak labels, the property of this technique is accepted but it is very vulnerable to the measure used to identify visual neighbours. New studies have employed learnable methods and weighted voting structures [18, 19, 23], or the images used for voting were more carefully selected [25]. In [17] the nearest neighbors are determined among different visual features with total of many measures [referred to as joint equal contribution (JEC)]. In Tag Propagation method (TagProp) [18] labels are calculated by considering a measured neighborhood balance of the labelled and unlabelled. Some methods are restricted for being biased against raising labels. Therefore, they drop in efficiency when faced to images with less related and unique labels.

Some studies have been worked on the extension of the neighbor voting algorithms. In [36] an image annotation model is presented in which similar clusters and multi-tag associations were extracted for each test image. To each nominee label a measure is calculated to reduce noisy tags. A visual graded neighbor voting has been introduced in [23] to enhance the efficiency of the neighbor voting method. The authors assigning a value to each vote which is synchronized with the correlation between the target image and the neighbor. In [22] also used image metadata to produce neighbor photos. The authors in this paper used deep learning to combine visual information between target image and its neighbors.

Also, some authors improved automatic image annotation approaches with tags refinement [15, 37, 38] and tags ranking [39, 40]. In [37] a method is proposed for label filtering, in which an invisible subject layer among images and labels is added. The label significance mechanism is introduced as a dot product between subject vector of the image and the subject vector of the label. The [38] model is based on matrix factorization. The image-label agreement matrix D as input and output led to the rebuilding of matrix $\hat{D}$. In [38], given that D is distorted by noise extracted from incomplete or over-personalized identifiers, reliable concept variable analysis is used to recover $\hat{D}$. They improve the solution by adding Laplacian regularizes on the similarity of images and tags, retaining a convex problem. In [40] a max-margin rummaged autonomy system is constructed for tag ranking.

The suggested approach in this article was adopted to address bias towards popular labels that reduce their efficiency while confronted with images with less similar and unique labels in the nearest neighbor image tagging. Our suggested method provides an alternative that is quick, scalable and makes high-quality suggestions.

3   Proposed Method

The value of the neighbor images labels are not same. Therefore, in this paper, to create a ranking process among images tags, a tree-based method is proposed. By using this method, the images which are more important for the input image, are placed at higher levels and their tags are also more important in tagging the input images.

The main parts of the proposed method are demonstrated in Fig. 1. First of all, before the image annotation process can be initiated and in offline stage, annotated image dataset is read and the feature extraction components are extracted from low-level features (color, texture, etc.) to feature vectors.

Fig. 1 The framework of the proposed method

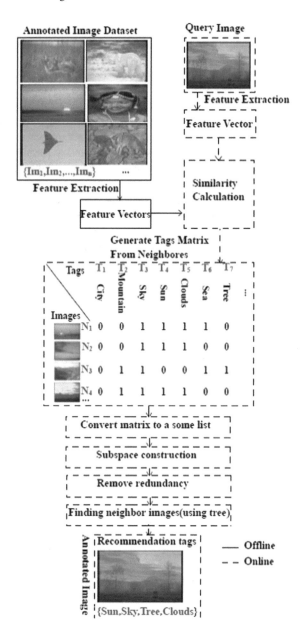

Definition 1 Find an image dataset, such as: $Img_1, Img_2, \ldots, Img_n$, where n is the number of images in image dataset φ. There are some t original labels, $T_1, T_2, \ldots, T_t$, and will find p top neighbor images (closest images), $N_1, N_2, \ldots, N_p$. Each top neighbor image contains several labels/tags. The term "top neighbor images" refers to the most similar images (the nearest neighbor lookup in a dataset) to the input

or query image. Let M be a $p \times t$ tag matrix of boolean values, t is the number of original labels and p is the number of the best neighbors.

Second, Once the person sends a target (test) image to the proposed method, the online stage begins, and the feature extraction part derives low-level (colour, texture, etc.) from the query image. Next, it is compared to every single training image. Similarity measures are computed in similarity calculation step in order to select candidate training images, and its neighbor images are retrieved to a tag matrix (M matrix with dimensions $p \times t$). Subsequently, various subspaces of relevant labels are generated and the subspaces of irrelevant labels are deleted from the lists. By processing tag subspaces, the neighbor images are recognized as elements in the next step. Then, a tree of images which are all neighbor images of the query image is constructed. Finally, recommendation labels to the input image are produced using the tree structure.

The following descriptions of these phases are mentioned in the following subsections.

3.1 Feature Extraction

Converting images into features is a vital phase for an image annotation model. This part makes feature from image. Local and global image features are used in the image annotation methods. Perhaps a pre-processing step such as clustering, extracting context, and segmentation will be performed prior feature extraction. Global descriptors are computed in all images whereas region descriptors are provided by image segmentation. Both Local and global descriptors are used in this paper to improve method efficiency. For more detail on the used features for this article, please see Sect. 4.4.

3.2 Similarity Calculation

Image similarity measures produce a quantitative evaluation of the similarity between two images. Euclidean measure is used as a metric of comparison in this paper. Euclidean similarity has many applications in information retrieval fields [41, 42].

Euclidean similarity is known as:

$$Sim(X, Y) = (\sum_{i=1}^{n} (\|X\| - \|Y\|^{h})^{\frac{1}{p}})$$

where h is regarded the normal variable and it is always $h \geq 1$. If h = 1 it is considered to be Manhattan distance, if h = 2 it is Euclidean and if $h = \infty$ it is Chebyshev.

$$Sim(X, Y) = \sqrt{\sum_{i=1}^{n} (X_i - Y_i)^2}$$

where X and Y are input image vector and image dataset vector, respectively. In this step, p top neighbor images (closest images), $N_1, N_2, \ldots, N_p$ with t labels/keywords to each p, $T_1, T_2, \ldots, T_t$ are found.

3.3 Generate Tag Matrix from Top Neighbors

In this step, a tag matrix is created. Let M be a $p \times t$ matrix of boolean values, p is the number of the best neighbors and t is the number of original labels/keywords. Anywhere M_{ij} element represents the tag i it is in the top neighbors j or not. Assignment matrix M is defined as

$$M_{ij} = \begin{cases} 1 & \text{if } tag\ i \in image\ j \\ 0 & \text{otherwise} \end{cases}$$

t value is fixed but p value can be change.

Later in this step; The Boolean matrix should show a list of relevant labels for each neighbor image. Therefore, irrelevant labels are ignored in the Boolean matrix (zero value labels). Doing this, dimensionality of matrix is reduced and only the information is saved which is important for the proposed method. For example, Fig. 2b demonstrates the converted version of the Boolean matrix that is shown in Fig. 2a. The goal of this phase is to decrease input label matrix sizes. The output data are showed by different lists of relevant labels.

Fig. 2 **a** Binary matrix.
b List of interesting tags

(a)

	T_1	T_2	T_3	T_4	T_5	T_6	T_7	T_8
N_1	0	1	1	0	0	0	0	0
N_2	1	0	1	0	0	1	0	0
N_3	0	1	0	1	0	0	0	0
N_4	0	1	1	0	0	0	1	0
N_5	0	1	1	1	0	0	0	0
N_6	1	1	1	1	0	0	0	0
N_7	0	0	0	0	1	1	1	1
N_8	0	1	0	0	0	1	0	0
N_9	0	0	1	0	1	1	1	0

(b)

N_1	T_2	T_3		
N_2	T_1	T_3	T_6	
N_3	T_2	T_4		
N_4	T_2	T_3	T_7	
N_5	T_2	T_3	T_4	
N_6	T_1	T_2	T_3	T_4
N_7	T_5	T_6	T_7	T_8
N_8	T_2	T_6		
N_9	T_3	T_5	T_6	T_7

3.4 Subspace Construction

The aim of this phase is to identify a subgroup of relevant labels that may be used to recognize a collection of related neighbors. We use a search method similar to the method developed by Vaizman et al. [41], Pickup et al. [42] is used. The search begins so as to compare each neighbor image with its sub-neighbor images (neighbor image N_i with neighbor images N_{i+1} to N_p). A local table is required to save local subsets for each adjacent image and its content is placed into a global table while the repetition for each adjacent image ends. For instance, if the search begins from neighbor image N_1, Then that will attempt to find the intersection between the photos of the neighbor N_1 and N_2. The result of the intersection for the instance displayed in Fig. 3 is tag T_3. After which, in this iteration, neighbor image N_1 is compared with neighbor image N_3, and the intersection T_2 is located in the local table. The intersections of neighbor image N_1 and neighbor image N_4, which is (T_2, T_3), are put in the local table. The process of intersection continues until images are compared. At the end of the intersection process for each adjacent image, the items in the local table are transmitted to the global table, which only accepts items not already show in the global table and removes the contents of the local table. When the search ends, 11 members as seen in Fig. 3. The pseudo code for seeking subspaces is as in as in the algorithm 1, and the pseudo code of the Find Interaction vector membership function is as in the algorithm 2.

Fig. 3 Final subspaces

1	T_3		
2	T_2		
3	T_2	T_3	
4	T_1	T_3	
5	T_3	T_6	
6	T_2	T_4	
7	T_7		
8	T_3	T_7	
9	T_2	T_3	T_4
10	T_6		
11	T_5	T_6	T_7

Algorithm. 1 Finding subspaces of interesting tags
1: **Input:** Listed matrix L and p number of rows.
2: **Output:** S: Subspaces
3: $local_table = \phi$, $global_table = \phi$
4: **for** j=0 to p **do**
5:     $local_table = \phi$
6:     row_j=L(j)
7:     **for** k=j+1 to p **do**
8:         row_k=L(k)
9:         I=**FindInteraction**(row_j,row_k)
10:         **put** I to $local_table$
11:     **end for**
12:     *move* $local_table$ entries to $global_table$
13: **end for**
14: S: $global_table$

Algorithm. 2 Find Interaction Function
1: **Input:** Two sorted string vectors (row_j,row_k).
2: **Output:** Subscription of vectors (S)
3: u=1
4: $I = \phi$
5: **for** i=0 to length of row_j **do**
6:     **for** c=0 to length of row_k **do**
7:         **if** $row_j[i] < row_k[c]$ **then**
8:             *Break;*
9:         **end if**
10:         **if** $row_j[i] = row_k[c]$ **then**
11:             $S[u]=row_j[i]$
12:             u++;
13:             *Break;*
14:         **end if**
15:     **end for**
16: **end for**

3.5 Delete Duplicate

A table of label subspaces was created in the previous phase. Several of the subspaces from that table are replicated in another subspaces that are considered duplicate. Deleting duplicate subspaces also allows the suggestion process to be quicker and easier. For example, Fig. 4a demonstrates the subspace T_3 is subsumed by (T_2, T_3), $(T_1, T_3), (T_3, T_6), (T_3, T_7)$ or (T_2, T_3, T_4) subspaces. This duplicate subspace is deleted in this phase.

Fig. 4 a Subspaces.
b Sorting subspaces based on the length of each subspace and removing the items used. **c** Ultimate items without overlapping

(a)				(b)				(c)			
1	T_3			1	T_2	T_3	$T_4{}^{\cdot}$	1	T_2	T_3	T_4
2	T_2			2	T_5	T_6	$T_7{}^{\cdot\cdot}$	2	T_5	T_6	T_7
3	T_2	T_3		3	T_2	$T_3{}^{\cdot}$		3	T_1	T_3	
4	T_1	T_3		4	T_1	T_3		4	T_3	T_6	
5	T_3	T_6		5	T_3	T_6		5	T_3	T_7	
6	T_2	T_4		6	T_2	$T_4{}^{\cdot}$					
7	T_7			7	T_3	T_7					
8	T_3	T_7		8	$T_3{}^{\cdot}$						
9	T_2	T_3	T_4	9	$T_2{}^{\cdot}$						
10	T_6			10	$T_7{}^{\cdot\cdot}$						
11	T_5	T_6	T_7	11	$T_6{}^{\cdot\cdot\cdot}$						

The first phase to delete unnecessary subspaces is to sort them depends on the number of labels used. Figure 4b demonstrates the sorted representation of Fig. 4a. In continue, conflicting subspaces with smaller sizes, as seen by the red color in Fig. 4b, are deleted Fig. 4c.

3.6 Finding Neighbor Images

Photos similar to the target image are identified in this step. For each subspace collection, neighbors that are involved in all tags of the subspace are inclusive in the neighbor list. For sample, neighbor 5 (N_5) and neighbor 6 (N_6) Are included in all elements of the subspace (2, 3, 4) (Fig. 5b).

After that, a tree represents neighbor images which are similar to the input image. The most related neighbor images are neighbor image 5 (N_5) and neighbor image 6 (N_6), according to the subspace neighbor image list. Thus, these images are positioned at the first level of the tree of the neighbor image. Defining the tree's second level one has to consider the images that overlap the first level images. For instance, image N_2 is similar to image N_6, then it is in the second level of the tree Fig. 5c. In

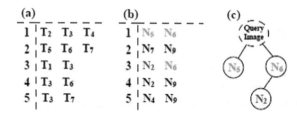

Fig. 5 a Ultimate subspaces. **b** Collection of photos that are included in the labels of each subspace and consider overlaps in collections of images of each item. **c** Tree of the neighbor image of the input image

the other words, N_2 added to the second level in order to N_2 and N_6 are on the 3rd row, and N_6 is located in the first priority.

3.7 Recommendation Tags

The lists of tags are created from the gained neighbors' images tree to the input image, as the final stage and to conclude the prior stage. The input is a representation of the present image dataset. All the adjacent images related to the target image are identified, and the appropriate neighbor images are rated based on the query scope. The most similar adjacent images are listed first, and the ranked process of the neighbor images dictates the overlap of the recommendations (first level). First, a set of neighbor images tags from neighbor images tree is created. Second, tags obtained for the tree level's priority are shown. Each image belongs to a level of the tree which is used for priority determining. Figure 6 is shown this process. For instance, tags (T_1, T_2, T_3, T_4) which have a higher priority than T_6, because of neighbor images 5 and 6 (N_5 and N_6) are located in the first level and neighbor image 2 (N_2) is located in the second level. The image tags with higher levels have higher priority compared with lower ones. All the tags are indicated in the example, however if the number of tags is changed from 5 to 4, then the tag number 6 (N_6) would be deleted since it only exists in image 2 (N_2) which is located at the tree's second level and it has lower priority than the other tags.

4 Experimental Results

The experimental results and various effects of the parameters will be explained in this section.

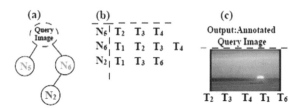

Fig. 6 **a** neighbor images tree. **b** Set of neighbor images tags. **c** Annotated query image

4.1 Experiment Setting

The detailed experiments were conducted using the Corel5k, IAPR TC12 and MIR Flicker datasets to demonstrate the efficacy of the proposed approach. For all three datasets, the number of images for nearest neighbors is supposed to be 50, and the number of labels for any query image is supposed to be 5, respectively. With higher number of images neighbors and tags, precision/recall can be increased while other measure will decrease. These statistics are a fair offer among precision and recall. The methods are programmed in MATLAB language and implemented with Intel Core i5 CPU 2.4 G and 4 G RAM on personal computer. Settings are the same in all experiments.

4.2 Datasets

Corel5k: Corel5k dataset is one of the common datasets for image annotation, consists of 5000 192 × 128 pixels of high and wide, tiny photos. Images were divided into 50 groups, and each group included 100 images which were generally similar. All 5000 images are split into a 4500-image training set and a 500-image test group. For label images 374 labels (keywords) are used. Distribution of 1 for 5 labels for each image is used [43].

    IAPR TC12: The collection was first used in 2006 through 2008 for the ImageClef Project. This collection is categorized into two categories consisting 17,665 for the training set and the test set with 1962 images. 291 labels have been used. That image received 1–23 labels, which is to indicate an average of 5.7 labels per image.

    MIR Flickr: This image dataset has been used in different image processing applications such as image retrieval and annotation. Images are downloaded from the social site of Flickr in this benchmark set [44]. The collection of MIR Flickr comprises 25,000 images for 38 names, with manual annotation.

4.3 Evaluation Metrics

Three metrics are used in this paper to test the annotation results efficiency: precision (P), recall (R) and F-score (F+). Annotation precision is the number of images that the label appropriately assigns divided by the total number of images that are automatically annotated by the label. Recall is the number of images that the label appropriately assigns divided by the total number of images that are annotated by the manual label. The average precision and recall on all positive-recall labels are used to test the following.

    Other metric is the F-measure, that is derived from a precision-recall comparison. F-measure is defined as:

$$F-measure(w) = \frac{2 \times P(w) \times R(w)}{P(w) + R(w)}$$

4.4 Features

Corel5k and IAPR TC12 datasets are used to experiment 15 various low-level features extracted by Guillaumin et al. [18]. These have one Gist feature [45], 6 global color histograms, and 8 descriptors of local visual bag-of-words. The suggested method uses estimation $\ell 1$ for global color feature (i.e. the Euclidean measure after translation is estimated to $\ell 1$ measure) and estimate measure $\chi 2$ for local bag-of-visual-words descriptor [13, 17, 46]. In a case similar to paper [47] various features including 1000D BoW and global feature, i.e. Color Histogram (64D) are used for MIR Flickr dataset.

4.5 Results

Firstly, outputs of the suggested method on Corel5k dataset are discussed. The suggested method compared to the below similar methods for image annotation:

- Cross-Media Relevance Model (CMRM) [48]: A generative method used for learning images semantics.
- Random Forest with Clustering-Particle Swarm Optimization (RFC-PSO) [49]: A framework based on class grades optimization using PSO, and random forest classifier.
- Joint equal contribution (JEC) [17]: A search-based approach that mixes visual features distances.
- Fast image tagging (FastTag) [46]: An approach with two simple linear mappings co-regulated in a joint curved loss function.
- Neighbor voting for social tag relevance (TagRel) [20]: A neighbor voting method for tag relevance exploration.
- Group Sparsity (GS) [13]: A framework based on groups sparsity and clustering from image features.

The mentioned algorithms are similar to our algorithm in aspect of the nearest-neighbor-based methods. It means that all algorithms use search-based image annotation and query image compared to every single training image in order to select candidate training images.

Figure 7 compares the model presented in this paper and past version of automated image annotation methods. It can be seen that the suggested model performs considerably better than all methods, with the exception of average recall and F-measure in

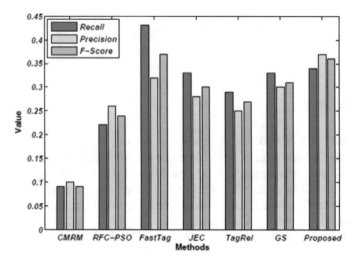

Fig. 7 Output of different annotation methods on Corel5k versus suggested method

the FastTag proposed [46]. It is reasonable, since our method focuses on the mining relevance tags and does not make models for tags.

Second, experimental results derived from the dataset IAPR TC12 will be discussed. Figure 8 compares the current methods dataset with the proposed method. Again, it is obvious that our models performed significantly much better than the current methods, except for average precision in recently proposed FastTag. It is sensible that the average keyword for each image in IAPR TC12 is more than Corel5k

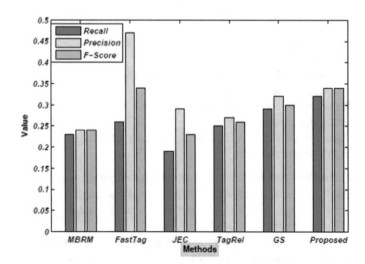

Fig. 8 Output of different annotation methods on IAPR TC12 versus suggested method

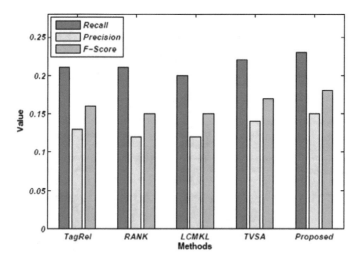

Fig. 9 Output of different annotation methods on MIR Flickr versus suggested method

(5.7 vs. 3.5) and our method focuses on the mining relevance tags and does not make models for single tags.

Finally, there will be reported experimental results collected from the MIR Flickr dataset. The proposed method is compared to TagRel [20], RANK [50], LCMKL [32] and TVSA [47]. That allows a true comparison among our methodology and other methods, since all strategies use the same visual features. Figure 9 demonstrates image annotation output on MIR flickr dataset. Similar to the results of other datasets, it demonstrates that the suggested model outperforms on average precision, recall and F-measure with the top 5 appropriate labels in contrast with the other models. The improvement generated by the approach described, though, is not as evident as the improvement of the other dataset. The major reason is that MIR Flickr's number of terms is only 38 and some such as bird-r1 and fly, or cloud-r1 and cloud are practically replicated.

Figure 10 displays several sample images with the effects of the suggested model of image annotation from the datasets Corel5 K, IAPR TC12 and MIR Flickr compared to human annotation in the context and TagRel. In all datasets five labels are used for each input image. In several photos, the images are very well identified by extra labels like "stadium" in the third image of the IAPR TC12 dataset in Fig. 10. Additionally, annotations provided by the presented model are more logical than TagRel.

Next, the results of the number of labels for any sample image on the presented method were analyzed by differing the number of neighbors. The average precision and recall of Corel5k dataset are displayed in Fig. 11, when the number of labels differs between 3 and 7 for any sample image. The average precision of the suggested approach rises as the number of labels rises as expected but the average recall reduces.

Image (Corel5k)				
Human Annotation	cars, tracks, turn, prototype	jet, plane, smoke, sky	field, foals, horses, mare	sky, sun, tree, water
Neighbor Voting Annotation(TagRel)	cars, street, turn, bear, tracks	sky, jet, water, plane, f-16	grass, field,tiger, horses, mare	sun, tree, sky water, clouds
Proposed Method Annotation	cars, tracks, grass, turn, prototype	sky, jet, plane, smoke, clouds	field, horses, mare, tree, foals	sky, tree, sun, couds, water
Image (IAPR TC12)				
Human Annotation	coast, shore	bed, bedcover, lamp, lawn, room, table,wall, window	court, man, player, short, tee-shirt, tennis,woman	cliff, lookout, hill, jersey, man, rope, sky, trouser
Neighbor Voting Annotation(TagRel)	cloud, coast, house, shore, tree	bed, curtain, front, lamp, room,	court, man, player, short,,tennis	cliff, helmet, hill, jersey, man
Proposed Method Annotation	cloud, coast, house, shore, sky	bed, room,wall, window, table	court, man, player, stadium,tennis	sky, cliff, man, hill jersey
Image (MIR Flickr)				
Human Annotation	animals, bird, bird_r1	plant_life, sky, tree	structures, transport	plant_life
Neighbor Voting Annotation(TagRel)	animals, people, bird indoor, sky	structures, plant_life, people, indoor, female	indoor, people, sky plant_life, structures	structures, plant_life, people, indoor, male
Proposed Method Annotation	animals, bird indoor, bird_rl, sky	plant_life, sky,indoor, tree, structures	structures, sky,transport plant_life, indoor	structures, plant_life, water, indoor, lake

Fig. 10 Corel5 K (first row), IAPR TC12 (second row), and MIR Flickr (third row) estimated labels versus human annotations for photos. The labels are estimated using the framework we suggest and the TagRel method

From Fig. 11 a trade-off between average precision and recall can be found while the number of labels for any sample image is increased.

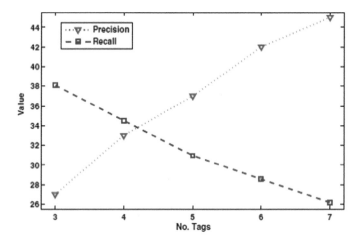

Fig. 11 Output of suggested model on Corel5k in varies labels

5 Conclusion

An efficient system for image annotation is introduced in this article. To display the image interest labels, the non-redundant subspaces are mined (extracted), then tree structure of neighbor images is drawn for the input image. Finally, a list of suggestions for labels is given for the input image. The suggested method is tested by contrasting it with other methods on some standard image annotation datasets. Compared with many similar techniques it shows better performance.

In the future works, someone might apply the proposed method to other datasets, other data structures and other visual attributes such as the features of deep neural networks. Also, it is an interesting subject to find a better approach for tag recommendations from the created neighbor image tree or improve the current approach. In addition, it is important to examine the application of distance metric learning methodology to discover more accurate visual neighbors.

References

1. S.U.N. Jun-ding, D.U. Juan, Review on automatic image semantic annotation techniques. Pattern Recognit. **45**(210128), 346–362 (2012)
2. A. Alzu'bi, A. Amira, N. Ramzan, Semantic content-based image retrieval: a comprehensive study. J. Vis. Commun. Image Represent., **32**, 20–54 (2015). https://doi.org/10.1016/j.jvcir. 2015.07.012
3. T. Deselaers, H. Müller, P. Clough, H. Ney, T.M. Lehmann, The CLEF 2005 automatic medical image annotation task. Int. J. Comput. Vis. **74**(1), 51–58 (2007). https://doi.org/10.1007/s11 263-006-0007-y

4. Y. Shin, Y. Kim, E.Y. Kim, Automatic textile image annotation by predicting emotional concepts from visual features. Image Vis. Comput. **28**(3), 526–537 (2010). https://doi.org/10.1016/j.ima vis.2009.08.009
5. C. Huang, F. Meng, W. Luo, S. Zhu, Bird breed classification and annotation using saliency based graphical model. J. Vis. Commun. Image Represent. **25**(6), 1299–1307 (2014). https:// doi.org/10.1016/j.jvcir.2014.05.002
6. G. Allampalli-Nagaraj, I. Bichindaritz, Automatic semantic indexing of medical images using a web ontology language for case-based image retrieval. Eng. Appl. Artif. Intell. **22**(1), 18–25 (2009). https://doi.org/10.1016/j.engappai.2008.04.018
7. A. Lotfi, V. Maihami, F. Yaghmaee, Wood image annotation using Gabor texture feature. Int. J. Mechatron. Electr. Comput. Technol. **4**, 1508–1523 (2014)
8. C. Lei, D. Liu, W. Li, Social diffusion analysis with common-interest model for image annotation. IEEE Trans. Multimed. **18**(4), 687–701 (2016). https://doi.org/10.1109/TMM.2015.247 7277
9. J. Liu, M. Li, Q. Liu, H. Lu, S. Ma, Image annotation via graph learning. Pattern Recognit. **42**(2), 218–228 (2009). https://doi.org/10.1016/j.patcog.2008.04.012
10. J.H. Su, C.L. Chou, C.Y. Lin, V.S. Tseng, Effective semantic annotation by image-to-concept distribution model. IEEE Trans. Multimed. **13**(3), 530–538 (2011). https://doi.org/10.1109/ TMM.2011.2129502
11. V. Maihami, F. Yaghmaee, A genetic-based prototyping for automatic image annotation. Comput. Electr. Eng. **70**, 400–412 (2018). https://doi.org/10.1016/j.compeleceng.2017.03.019
12. D. Zhang, M. Monirul Islam, G. Lu, Structural image retrieval using automatic image annotation and region based inverted file. J. Vis. Commun. Image Represent. **24**(7), 1087–1098 (2013). https://doi.org/10.1016/j.jvcir.2013.07.004
13. S. Zhang, J. Huang, Automatic image annotation and retrieval using group sparsity. Syst. Man **42**(3), 838–849 (2012). https://doi.org/10.1109/tsmcb.2011.2179533
14. Y. Yang, Z. Huang, Y. Yang, J. Liu, H.T. Shen, J. Luo, Local image tagging via graph regularized joint group sparsity. Pattern Recognit. **46**(5), 1358–1368 (2013). https://doi.org/10.1016/j.pat cog.2012.10.026
15. L. Ballan, M. Bertini, G. Serra, A. Del Bimbo, A data-driven approach for tag refinement and localization in web videos. Comput. Vis. Image Underst. **140**, 58–67 (2015). https://doi.org/ 10.1016/J.CVIU.2015.05.009
16. Z. Chen, Z. Chi, H. Fu, D. Feng, Multi-instance multi-label image classification: a neural approach. Neurocomputing **99**, 298–306 (2013)
17. A. Makadia, V. Pavlovic, S. Kumar, A new baselines for image annotation. Int. J. Comput. Vis. **90**, 88–105 (2010)
18. M. Guillaumin, T. Mensink, J. Verbeek, C. Schmid, C. Schmid TagProp, TagProp: Discriminative metric learning in nearest neighbor models for image auto-annotation: Discriminative metric learning in nearest neighbor models for image auto-annotation, in *IEEE Computer society* (2009), pp. 309–316. https://doi.org/10.1109/iccv.2009.5459266
19. C. Cui, J. Shen, J. Ma, T. Lian, Social tag relevance learning via ranking-oriented neighbor voting. Multimed. Tools Appl. **76**(6), 8831–8857 (2017). https://doi.org/10.1007/s11042-016-3512-1
20. X. Li, C.G.M. Snoek, M. Worring, Learning social tag relevance by neighbor voting. IEEE Trans. Multimed. **11**(7), 1310–1322 (2009). https://doi.org/10.1109/TMM.2009.2030598
21. X. Li, T. Uricchio, L. Ballan, M. Bertini, C. G. M. Snoek, A. Del Bimbo, Socializing the semantic gap: a comparative survey on image tag assignment, refinement and retrieval, in *ACM Computer Survey Preprint. arXiv1503.08248* (2016) https://doi.org/10.1145/2906152
22. J. Johnson, L. Ballan, L. Fei-Fei, Love thy neighbors: Image annotation by exploiting image metadata, in *Proceedings of the IEEE International Conference on Computer Vision*, vol. 2015 Inter (2015) pp. 4624–4632, https://doi.org/10.1109/iccv.2015.525
23. S. Lee, W. De Neve, Y.M. Ro, Visually weighted neighbor voting for image tag relevance learning. Multimed. Tools Appl. **72**(2), 1363–1386 (2014). https://doi.org/10.1007/s11042-013-1439-3

24. Y. Verma C.V. Jawahar, Image annotation using metric learning in semantic neighbourhoods, in *Computer Vision–ECCV 2012* (Springer, 2012), pp. 836–849
25. A. Yu, K. Grauman, Predicting useful neighborhoods for lazy local learning, in *Advances in Neural Information Processing Systems* (2014), pp. 1916–1924
26. F. Tian, X. Shen, F. Shang, Automatic image annotation with real-world community contributed data set. Multimed. Syst. **25**(5), 463–474 (2019). https://doi.org/10.1007/s00530-017-0548-7
27. V. Maihami, F. Yaghmaee, Automatic image annotation using community detection in neighbor images. Phys. A Stat. Mech. Appl. **507**, 123–132 (2018). https://doi.org/10.1016/j.physa.2018.05.028
28. Y. Ma, Y. Liu, Q. Xie, L. Li, CNN-feature based automatic image annotation method. Multimed. Tools Appl. **78**(3), 3767–3780 (2019). https://doi.org/10.1007/s11042-018-6038-x
29. V. Maihami, F. Yaghmaee, A review on the application of structured sparse representation at image annotation. Artif. Intell. Rev. **48**(3), 331–348 (2017). https://doi.org/10.1007/s10462-016-9502-x
30. Q. Cheng, Q. Zhang, P. Fu, C. Tu, S. Li, A survey and analysis on automatic image annotation, in *Pattern Recognition* (2018)
31. J. Chen, D. Wang, I. Xie, Q. Lu, Image annotation tactics: transitions, strategies and efficiency. Inf. Process. Manag. **54**(6), 985–1001 (2018). https://doi.org/10.1016/j.ipm.2018.06.009
32. Y. Gu, X. Qian, Q. Li, M. Wang, R. Hong, Q. Tian, Image annotation by latent community detection and multikernel learning. IEEE Trans. Image Process. **24**(11), 3450–3463 (2015). https://doi.org/10.1109/TIP.2015.2443501
33. J. Huang, H. Liu, J. Shen, S. Yan, Towards efficient sparse coding for scalable image annotation, in *Proceedings of the 21st ACM International Conference on Multimedia* (2013), pp. 947–956. https://doi.org/10.1145/2502081.2502127
34. P.V. Bengio, Y. Aaron Courville, Representation learning: a review and new perspectives. Pattern Anal. Mach. Intell. IEEE Trans. **35**(8), 1798–1828 (2013)
35. A.G. Howard, Some improvements on deep convolutional neural network based image classification. *arXiv Prepr.* (2013)
36. Y. Yang, Z. Huang, H.T. Shen, X. Zhou, Mining multi-tag association for image tagging. World Wide Web **14**(2), 133–156 (2011). https://doi.org/10.1007/s11280-010-0099-8
37. J. Wang, J. Zhou, H. Xu, T. Mei, X.S. Hua, S. Li, Image tag refinement by regularized latent Dirichlet allocation. Comput. Vis. Image Underst. **124**, 61–70 (2014). https://doi.org/10.1016/j.cviu.2014.02.011
38. Y. Zhu, G. Yan, S. Ma, Image tag refinement towards low-rank, content-tag prior and error sparsity, in *Proceedings of the International Conference on Multimedia* (2010)
39. D. Liu, X.-S. Hua, L. Yang, M. Wang, H.-J. Zhang, Tag ranking, in *Proceedings of the 18th International Conference on World Wide Web—WWW '09* (2009), p. 351. https://doi.org/10.1145/1526709.1526757
40. G. Lan, T. Mori, A max-margin riffled independence model for image tag ranking, in *Proceedings of the IEEE Conference on Computer Vision and Pattern Recognition* (2013), pp. 3103–3110
41. G. Vaizman, Y. McFee, B. Lanckriet, Codebook-based audio feature representation for music information retrieval. IEEE/ACM Trans. Audio, Speech Lang. Process. **22**(10), 1483–1493 (2014)
42. R.R. Pickup, D. Sun, X. Rosin, P.L. Martin, Euclidean-distance-based canonical forms for non-rigid 3D shape retrieval. Pattern Recognit. **48**(8), 2500–2512 (2015)
43. P. Duygulu, K. Barnard, J.F.G. de Freitas, D.A. Forsyth, Object recognition as machine translation: learning a lexicon for a fixed image vocabulary, in *Computer Vision ECCV 2002* (Springer, 2002), pp. 97–112
44. M.J. Huiskes, B. Thomee, M.S. Lew, New trends and ideas in visual concept detection, in *Proceedings of the International Conference on Multimedia Information Retrieval—MIR '10* (2010), p. 527. https://doi.org/10.1145/1743384.1743475
45. A. Oliva, A. Torralba, Modeling the shape of the scene: a holistic representation of the spatial envelope. Int. J. Comput. Vis. **42**(3), 145–175 (2001)

46. M. Chen, A. Zheng, K. Weinberger, Fast image tagging, in *Proceedings of the 30th International Conference on Machine Learning (ICML-13)*,vol. 28 (2013), pp. 1274–1282
47. Y. Gu, H. Xue, J. Yang, Cross-modal saliency correlation for image annotation. Neural Process. Lett. **45**(3), 777–789 (2017). https://doi.org/10.1007/s11063-016-9511-4
48. J. Jeon, V. Lavrenko, R. Manmatha, Automatic image annotation and retrieval using cross-media relevance models, in *26th Annual International ACM SIGIR Conference on Research and Development in Information Retrieval* (2003), pp. 119–126
49. N. El-Bendary, T. Kim, A.E. Hassanien, M. Sami, Automatic image annotation approach based on optimization of classes scores. Computing **96**(5), 381–402 (2013). https://doi.org/10.1007/s00607-013-0342-0
50. M. Wang, Tag Ranking, in *Proceedings of the 18th International Conference on World Wide Web* (2009), pp. 351–360

A Survey: Implemented Architectures of 3D Convolutional Neural Networks

Rudra Narayan Mishra, Utkarsh Dixit, and Ritesh Mishra

Abstract Convolutional Neural Networks are those types of deep neural networks that have shown great performance on a varied set of benchmarks. This powerful learning ability of a convolutional neural network is mainly due to the use of several hidden layers that can automatically extract all the important features from the data. Due to the availability of datasets in large amount and discovery for better hardware units has extended the research in CNNs to a great extent, and due to such extensive research work, a large number of architectures have been reported and implemented on real-world data. The introduction of 3D sensors encouraged pursuing research in visionary aspect of computers for many real-world application areas including virtual-reality, augmented-reality and medical imaging. Recently, many 2D and 3D classification methods depend on CNNs, which have a reliable resource in extracting features. But, they cannot find out all the dimensional relationships between features due to the max-pooling layers, and they require a vast amount of data for training. In this paper, we survey on different implementations of 3D Convolutional neural networks and their respective accuracies for different datasets. We then compare all the architectures to find which one is most suitable to perform flexibly on CBCT scanned images.

Keywords CNN · CBCT · 2D · 3D

1 Introduction

Many machine learning algorithms ensures that computers acquire intelligence by establishing different relationships among those data attributes and makes them learn from those relationships. Different Machine learning algorithms have been developed for attaining the desired superiority to achieve the human-level-satisfaction [1–6].

R. N. Mishra (✉) · U. Dixit · R. Mishra
School of Computer Engineering, Kalinga Institute of Industrial Technology (KIIT) deemed to be University, Bhubaneswar, Odisha, India
e-mail: 1705441@kiit.ac.in

© Springer Nature Switzerland AG 2020
P. K. Mallick et al. (eds.), *Cognitive Computing in Human Cognition*,
Learning and Analytics in Intelligent Systems 17,
https://doi.org/10.1007/978-3-030-48118-6_3

This challenging nature of such algorithms gives rise to an exclusive class known as Convolutional Neural Network.

Object recognition algorithm that try to identify certain objects in an image or videos, normally depends on comparing, training, or pattern recognition algorithms using appearance or feature based techniques. The application of this technology related to such fields is increasing slowly but surely. Moreover, the constructiveness of 3D object recognition is proliferating as researchers are finding new algorithms, models and approaches and implementing them [7]. The applications for which 3D object recognition and analysis technology is used are:

1. **Manufacturing Industry**: 3D object recognition technology can make tasks like manufacturing or maintenance more efficient in the aerospace, automotive and machine industries. The advances in 3D object recognition technology is making robots more intelligent and autonomous as issues such as recognition of handing objects and obstacles during navigation are being addressed [7].
2. **Video surveillance**: The usage of 3D object recognition technology in this area includes tracking people and vehicles, segmenting moving crowds into individuals, face analysis and recognition, detecting events and behaviours of interest and scene understanding [7].
3. **Medical and Health care**: Deep learning and 3D object recognition are being widely used in computer-aided diagnosis systems in the medical industry. Various applications include analysis of 3D images from CT scans, digital microscopy, and related fields [7].

In the recent past, many surveys were conducted on deep CNNs that elaborate on the basic components of CNN [8–10]. Many surveys have reviewed the famous architectures. Whilst, in this survey, we explore different implementations of 3D CNN algorithms on different 3D datasets in order to find the versatile method for classifying scanned CBCT images.

The various 3D CNN architectures discussed in this survey are CapsNet, 3D U-Net, V-Net, DenseNet and PointNet [7–11]. This chapter is organized as follows:

Section 1 states how to manipulate different input data as per our Region of Interest (ROI).

Section 2 discusses all the above-mentioned methods in a brief manner.

Section 3 will provide all the compared results in a tabular manner so as to find which one is a better model for 3D medical images.

Section 4 will provide the conclusion statement of this survey paper.

2 Input Data Manipulation

The most difficult challenge which is faced usually is to manipulate the input of the 3D images into a form that is more suitable for the designed architecture of the CNN [12]. For 2D CNN, the image is usually 2D in nature that can be in RGB format or in grayscale which is a n × n matrix with input channel r that corresponds to the different layers of the image. For transferring the 3D CNN into the usual CNN format can be achieved if we visualise as the 3rd dimension, thus, we can achieve that format by changing the input dimension of n × n × r to m × m × m. This requires a small modification to the architecture of the CNN rather than changing the whole input [13].

In 3D environmental mapping for aerial vehicles, we use volumetric occupancy grid that represents the present state of an environment through a matrix of randomly initialized numbers where each number corresponds to a voxel, simultaneously maintaining a probabilistic estimate of the occupancy in each voxel based on the incoming range sensor data and previous input provided [13]. The benefits of using such grids are that they allow an efficient estimation of free, occupied and unknown space of the working environment, even if using multiple sensory inputs that do not have the same origins or occurrence at the same time instance, thus discretizing the data points from all the floating points in the 3D space into more evenly spaced voxels in the 3D grid [13].

3   Architectures for 3D CNN

By adding several layers of neurons, we can improvise the algorithm's extracting power. Thus, the depth of the neural network has played a major role in the success of supervised learning and training. In this section, we are going to discuss various 3D CNN models that are implemented by different PhD scholars in their thesis papers.

3.1   3D U-Net Architecture

The training of any deep neural networks usually requests for a large training data with proper labels, which is very rare in the context of medical image processing. But, by using data augmentation to available training images, we can get many flexible deformations and disorientation in the images, and U-net is proposed to learn the augmented data set [8]. U-net is based on Fully Connected Network, but it extends the FCN architecture by changing the upsampling part of the data such that it can also possess many feature channels. U-net structure has both a contracting side as in any classic convolution network and an expanding side (deconvolution) that includes upsampling part, due to which it forms a U-shaped architecture [8].

3D-Unet is the 3D version of the original U-net that was proposed for the reason that medical images usually have to be in the form of 3D volumetric type such as Nifti, Dicom, instead of 2D images [8]. The high quality illustrated training data is even harder to acquire as it involves a time-consuming and tedious job of labelling the region of interest slice by slice. 3D-Unet takes 3D images as input and is trained to learn from inadequate illustrated images and predict the dense 3D segmentation. A

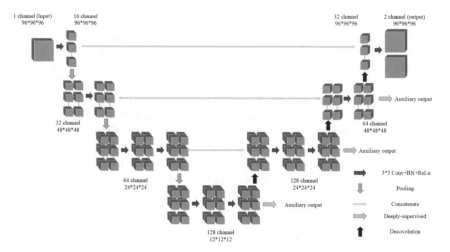

Fig. 1 Example of 3D-Unet structure

trained 3D-Unet model can also be applied to a new volumetric image and predicts its output dense 3D segmentation [8]. Figure 1 shows an example of 3D-Unet structure.

In [14], MRI images of the human spine and its label files for vertebral bodies are used as datasets. The image format was Neuroimaging Informatics Technology Initiative (Nifti).

The model received Nifti images comprising of 6 patients for training and 2 sets of images for prediction purposes. The result shows that the model predicting area was smaller and also the prediction had some noisy points.

3.2   V-Net Architecture

U-Net based CNNs have already proven to be a good solution for 3D CNN architecture. Despite this, they have a fundamental design flaw: the lack of volumetric context, as they ignore differences along the z-axis. This essentially means that each slice is segmented as an individual image, which could make it harder to segment smaller patches, like a tumour, and reduce performance. Özgün et al. address this issue with V-Net [9]. V-Net is an architecture based on U-Net, where instead of using an image as input, it expects a volumetric shape [9].

The architecture of V-Net is given by Fig. 2.

At first, the input gets compressed to a lower resolution and then it gets decompressed to have the original input image back. Each stage on the left part includes one to three convolutional layers. The convolutions use volumetric kernels having a size of $5 \times 5 \times 5$ voxels. Hence, the convolved feature maps size is reduced by half, doubling the number of feature channel at each stage of the V-Net compression

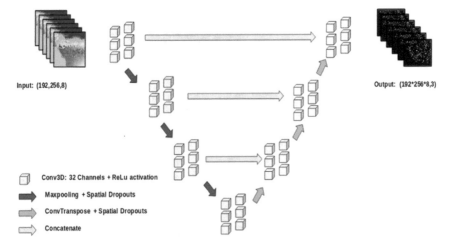

Fig. 2 Example of 3D-vnet structure

path, that is similar to pooling layers. PReLU is used as the activation function in the V-Net architecture.

In the right part, the network extracts features and deconvolves the input size by using one to three convolution layers having $5 \times 5 \times 5$ kernel. The two features maps that are having $1 \times 1 \times 1$ kernel size, computed by the very last layer produces the outputs of the same size as the input volume. These two layers segments the foreground and the background by applying soft-max activation function at each voxel.

ISBI 2017 Liver Tumour Segmentation Challenge Challenge1 (LiTS) dataset was used for experimentation purposes [9]. It was comprised of 200 contrast-enhanced abdominal CT scans in the Nifti format, from 6 medical centres, where, 130 was meant for training, and 70 was meant for testing. Each scan, represented by a volume, has a harmonized segmented file as base truth, containing a rough liver segmentation as Label 1, and a correspondent expert's lesion segmentation as Label 2, as seen in the given Figs. 3 and 4 [9].

Tridimensional processing of a CT scan is very computationally expensive. Because of rescaling, some data was lost due to memory insufficiency. This, unfortunately, means that Volumetric CNNs still cannot compete with a normal 2D approach, assuming they use full resolution volumes. But, when comparing results with 2D CNNs using the downscaled dataset, 3D CNNs excelled them [9]. This is evidence of the importance of the z-axis when segmenting CT scans [9].

Fig. 3 CT volume [9]

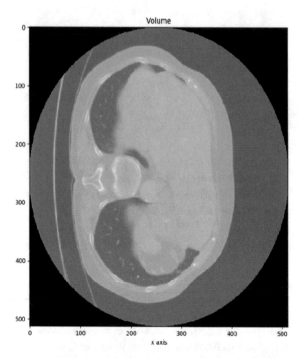

Fig. 4 Liver segmentation
with two lesions [9]

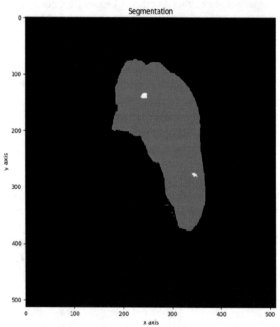

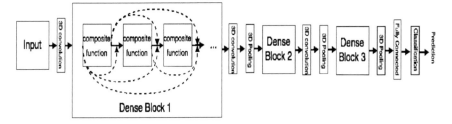

Fig. 5 Dense block architecture

3.3   DenseNet Architecture

3D-DenseNet is a deep network that adds some extra convolution, pooling layers in DenseNet which makes it work on any discrimination task [10]. 3D-DenseNet is comprised of four basic layers, namely, the 3D convolution layer, 3D pooling layer, rectified linear units (ReLU) layer, and the fully connected layer of a 3D CNN [10].

3D pooling is an essential operation that helps to reduce the size of the feature-maps. To use 3D pooling in our architecture we take the multiple densely connected dense blocks idea from the Fig. 5.

For the experiments, 3D-Densenet has three dense blocks with the same number of layers and used a 3D convolution layer with kernel size $3 \times 3 \times 3$ on the input images before entering the first dense block. The output channel for this convolution layer is twice the growth rate of 3D-DenseNet [10]. For each convolution layer inside the dense block, each side of the input is zero-padded by one pixel to keep a fixed-size feature-map. The transition layer between two dense blocks consists of a $1 \times 1 \times 1$ convolution layer and a $d \times 2 \times 2$ average pooling layer, where d is 1 on the first transition layer and 2 on the second transition layer. Inside the classification layer, a global average pooling is used, followed by a softmax classifier [10]. The table below presents the general 3D-DenseNet version of our model. Each dense block has one more $1 \times 1 \times 1$ Conv operation, compared with that of the general 3D-DenseNet in order to mitigate overfitting and reduce the training parameters (Fig. 6).

The MERL dataset was chosen for the testing set. In addition to that, HOG people detector was used on KTH to extract the bounding box of a person as the input of the network [10]. Image examples can be found in Figs. 7 and 8.

It was trained with the 3D-DenseNets with depths of 40 and growth rates of 24. The main results on the MERL shopping datasets were found to be 84.32% [10].

3.4   PointNet Architecture

PointNet is a deep neural network where the 3D point cloud is directly consumed which provides a unified approach for tasks such as classification and segmentation [11]. The important feature of this network is robustness with respect to the input

Layers	Output Size	3D-DenseNet-BC-20(k=12)	3D-DenseNet-BC-40(k=12)
3D Convolution	16 * 100 * 100	$3 * 3 * 3$ conv	
Dense Block 1	16 * 100 * 100	$\begin{bmatrix} 1*1*1 & conv \\ 3*3*3 & conv \end{bmatrix} * 2$	$\begin{bmatrix} 1*1*1 & conv \\ 3*3*3 & conv \end{bmatrix} * 6$
Transition Layer 1	16 * 100 * 100	$1 * 1 * 1$ conv	
	16 * 50 * 50	$1 * 2 * 2$ average pool	
Dense Block 2	16 * 50 * 50	$\begin{bmatrix} 1*1*1 & conv \\ 3*3*3 & conv \end{bmatrix} * 2$	$\begin{bmatrix} 1*1*1 & conv \\ 3*3*3 & conv \end{bmatrix} * 6$
Transition Layer 2	16 * 50 * 50	$1 * 1 * 1$ conv	
	8 * 25 * 25	$2 * 2 * 2$ average pool	
Dense Block 3	8 * 25 * 25	$\begin{bmatrix} 1*1*1 & conv \\ 3*3*3 & conv \end{bmatrix} * 2$	$\begin{bmatrix} 1*1*1 & conv \\ 3*3*3 & conv \end{bmatrix} * 6$
Classification Layer	1 * 1 * 1	$8 * 25 * 25$ average pool	
		fully connected	
		classification and softmax	

Fig. 6 DenseNet layering

Fig. 7 MERL dataset [10]

Fig. 8 Extracting the images [10]

video clip with length 1 second																									25 Images

apprehension and alteration. The network generally learns to encapsulate a shape by a scattered set of points. It also has a data-dependent dimensional transformer network that tends to authorize the data by applying a rigid transformation such that every point of the data will transform independently [11].

The input layer takes in a point cloud with n points, which is declared at the time of training graph declaration. Once n is fixed, each of these n points is treated as a 2D input and the input vector resembles the following dimensions—batch size * n * 1 * 3. The initial (per-point) feature extraction is realized with the 2D convolutions, where 3 channels are actually the x; y; z coordinates of that specific point. This is also termed as a shared- multi-layer perceptron in the paper. The details of shared multi-layered perceptrons are presented in a detailed visualization. Finally, a three-layer fully-connected network is used to map the global feature vector to k output classification scores.

Transformation Invariance: The classification (and segmentation) of an object should be invariant to certain geometric transformations (e.g., rotation). Motivated by Spatial Transformer Networks (STNs), the "input transform" and "feature transform" are modular sub-networks that seek to provide pose normalization for a given input [11] (Fig. 9).

The architecture is given below:

Berkeley MHAD has a total of 137,684 depth images acquired by each of the two Kinect cameras used in their experiment, which are synchronized for consistency [11].

The dataset was split into randomly selected 25% of the MHAD dataset as our test set, and keeping the rest for training. Accuracy Score was found to be with the mean error of 3.92 and an accuracy of 90% [11].

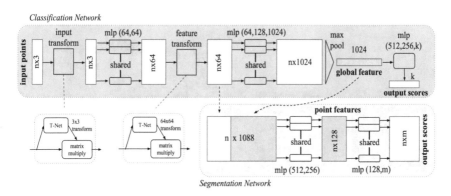

Fig. 9 PointNet architecture

3.5 CapsNet Architecture

CNNs have several deficiencies. Firstly, CNNs does not verify the relative positions of all the aspects in the data. Secondly, they need a large amount of data to hypothesize the classifier. Thirdly, it is believed that CNNs are not a good representation of the human vision system [7]. Capsule network (CapsNet) tries to overcome these shortcomings of CNNs. The main feature of CapsNet is that the neurons are reciprocated with capsules. A capsule is a container of several neurons that are represented by stacked layers of features. A CapsNet solves the issues of CNNs by augmenting the same image. Hence, less training data does not get bothered by the CapsNet. A group of neurons that does backend operations as a whole on inputs given to them and wrapping up the output in a vector for a single entity is known as a capsule [7].

CapsNet has a deep architecture with two primary capsule layers. The first layer is a convolutional layer with 64 filters of kernel size 7 and stride 1. Activation used is leaky ReLU. Dropout with probability to keep = 0.4 is used after the convolutional layer and batch normalization is used. The input to the first primary caps is 18 and the input to second primary caps is 12. The squashing function is used in the primary capsules to squash the capsule output to less than 1 for long vectors and 0 for short vectors. This proposed architecture is described in Fig. 10.

For experimentation, the Princeton ModelNet project was used that collected 3D CAD models of most common objects [7]. In the given architecture model, the 10-class subset of the full dataset was used. The classes in the modelnet10 dataset are bathtub, bed, chair, desk, dresser, monitor, nightstand, sofa, table, and toilet (Fig. 11).

For the experiment, 40% of the data was used to train the data. Since the model performed well on 40% training data, extra 5% split was given to training data and

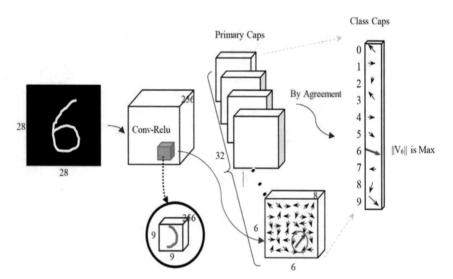

Fig. 10 CapsNet architecture

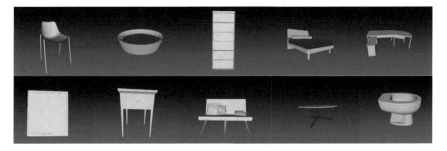

Fig. 11 ModelNet10 dataset [7]

15% split to perform experiments on the model. Accuracy found on the given dataset (ModelNet10) was *91.31%* using 3D CapsNet [7].

4   Results

Training of 3D CNN is unquestionably enhanced during the years by the exploitation of depth and other structural modifications. Observation of the recent literature has shown that the main boost in the performance of 3D CNNs has been accomplished by the replacement of the convolutional layer structure with small-scale partitions.

The below table shows the comparison of all the architectures we have discussed:

Architecture	Dataset used	Accuracy gained (%)
3D U-Net	MRI of human spine	92
3D V-Net	LiTS	92
DenseNet	MERL shopping	84.32
PointNet	MHAD dataset	90
CapsNet	ModelNet10	91

5   Conclusions

Unco progress has been made by 3D CNN, since the year 2013 when it was recognized as a better-modernised model of CNN specifically in the tasks which were related to vision and thus it encouraged and revived the interest in scientists in CNN. In the context of this, assorted research works have been conducted to enhance the performance of CNN on tasks related to vision mainly during the year 2015 and successive years to follow. This paper reviews evolution in the field of 3D CNN architecture based on the usage, design patterns and thus has found a suitable fit architecture for now.

References

1. O. Chapelle, *Support vector machines for image classification* (1998)
2. D. G. Lowe, *Object recognition from local scale-invariant features* (1999)
3. H. Bay, A. Ess, T. Tuytelaars, L. Van Gool, Speeded-up robust features (SURF) (2008)
4. N. Dalal, W. Triggs, Histograms of oriented gradients for human detection (2004)
5. T. Ojala, M. Pietikäinen, and D. Harwood, "A comparative study of texture measures with classification based on feature distributions," 1996
6. M. Heikkilä, M. Pietikäinen, C. Schmid, Description of Interest Regions with Local Binary Patterns (2009)
7. A. Ahmad, Object Recognition in 3D Data Using Capsules (2018)
8. B. Fan, Selective compression of medical images via intelligent segmentation and 3D-SPIHT coding (2018)
9. P. D. da Cunha Amorim, 3D convolutional neural network for liver tumor segmentation (2018)
10. D. Gu, 3D densely connected convolutional network for the recognition of human shopping actions (2017)
11. A. Ali, 3D human pose estimation (2019)
12. K. Schauwecker, A. Zell, Robust and efficient volumetric occupancy mapping with an application to stereo vision (2014)
13. A. Bender, E.M. Þorsteinsso, Classification of voxelized LiDAR point cloud data using 3D convolutional neural networks (2016)
14. J. Bouvrie, 1 introduction notes on convolutional neural networks (2006)

An Approach for Detection of Dust on Solar Panels Using CNN from RGB Dust Image to Predict Power Loss

Ritu Maity, Md. Shamaun Alam, and Asutosh Pati

Abstract Energy and Environment are important aspects of today's world. The solar Photovoltaic (PV) has become one of the emerging renewable energy source which has revolutionized energy sector across the world. In recent era growing number of solar farm are not free from maintenance and operation challenges. Environment induced dust on solar panel hampers power generation at large. This paper focuses on CNN based approach to detect dust on solar panel and predicted the power loss due to dust accumulation. We have taken RGB image of solar panel from our experimental setup and predicted power loss due to dust accumulation on solar panel.

Keywords Dust · Solar panel · CNN · Lenet model · RGB

1 Introduction

In recent era energy related aspects are becoming main area of concern. One of the popular renewable and clean energy source is solar energy. Power generation from solar energy the most effective method is by solar cell. For efficient conversion of solar energy, solar cell should be effectively handled and maintained. The major challenge in solar cell maintenance are various environmental issues [1]. Dust deposition on solar panel is also one of the major challenges as it leads to considerable loss in power generation. An effective intelligent detection system can improve solar farm operation and maintenance [2].

Our objective is to develop a system which can predict the amount of power loss due to dust deposition by using CNN Lenet based model. We have tried to present a simple CNN model which can predict the power loss as per the images of dusty solar panel.

R. Maity (✉) · Md. Shamaun Alam · A. Pati
Department of Mechanical, CTTC, Bhubaneswar, India
e-mail: maity.ritu07@gmail.com

© Springer Nature Switzerland AG 2020　　　　　　　　　　　　　　　41
P. K. Mallick et al. (eds.), *Cognitive Computing in Human Cognition*,
Learning and Analytics in Intelligent Systems 17,
https://doi.org/10.1007/978-3-030-48118-6_4

1.1 Solar Panel

Solar panels are used to generate direct current electricity by absorbing solar irradiation reaching its surface. Solar cell are available in different voltages and wattages. Solar panel modules consists of array of solar cells which generate renewable energy sources in various fields [3]. Efficiency of solar panel depends on maximum voltage generated, temperature, irradiation and environmental factors.

1.2 Need to Remove Dust on Solar Panel

Dust accumulation in solar panel is a major issue faced in field of renewable energy sector. Sun's irradiance is obstructed from reaching solar panel due to dust deposition on the panel. It minimizes photovoltaic energy generation by 5-20% in an average [4]. There are number of factors which determine the dusting effect on solar panel. The placement of solar panel (i.e. horizontal location accumulates more dust), surface of panel, wind direction and other geographical conditions. For generating maximum output from solar panel there should not be any dust particle which obstruct the solar irradiance from reaching the solar panel. The solar panels are placed horizontal to direction of solar irradiation facing upwards, for which they are exposed to dusting from surrounding environment [5]. The effectiveness and performance of solar panel decreases due to dust accumulation on the solar cell. For which it is important to know that how much dust has been accumulated and it leads to how much decrease in solar panel efficiency [6].

1.3 Challenges

Dust deposition is a greatest problem faced by solar farms which have drastically reduced performance of solar cell [7]. The extent of reduction in performance depends on the dust particle size and quantity of dust deposited on solar panel. The loss in power output of solar panel varies directly with mass of dust deposition and varies inversely with size of dust particles as smaller particles can considerably obstruct the path of solar ray's from reaching the surface of solar panel. The dust deposition on solar panel can include sand particle, red soil, white cement etc. [8]. Most of the researchers [2, 6, 8] discuss on effect of dust deposition on the surface of solar panel. It was proved that power efficiency of solar panel decreased considerably with deposition of dust which can be of different types from various research conducted in this field [8]. Deep solar eye [2] researcher had carried out convolutional neural network to predict power loss by using Impact net method. The dust on solar panel can be detected from RGB image of solar panel using automatic visual inspection system. The main challenge in using CNN approach to detect dust on solar panel is lack of labeled datasets. In image classification, labelling and detecting location of the required object is tedious task Our proposed approach consists of simple CNN.

Lenet model architecture which needs only solar panel images with loss in power output as labels for training keeping irradiance as constant.

2 Convolutional Neural Network

Deep learning is the emerging technology in field of smart computing. For image datasets, CNN is the most effective technique in deep learning for developing prediction models. CNN is a deep learning model inspired from visual cortex in human brain and it can learn automatically and adapt to different dimension of features from low to high level [9]. The architecture of CNN is such that it consists of basically three layers i.e. convolutional layer, pooling layer and fully connected layer that are stacked together to form different architecture. Convolutional layer perform feature extraction by passing a stride of 5 × 5 to extract useful information from the image and then passing it through Relu layer which removes all negative values. Then image is passed through a pooling layer where we shrink the image stack into smaller size by passing a stride of 2 × 2. Then it is passed through fully connected layer takes the extracted features to map the final output. The model is optimized using back propagation and gradient descent algorithms [9].

2.1 CNN Architecture

There are different types of CNN architecture. In this project we have used LeNet model. The LeNet architecture is an excellent "first architecture" for Convolutional Neural Networks. As shown in Fig. 1 this model comprises of two convolutional layer and two pooling layer and then images are passed through fully connected layer to get final output

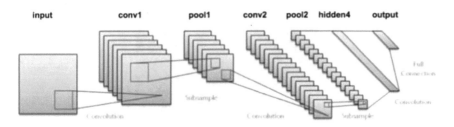

Fig. 1 Pictorial representation of a CNN Lenet model Architecture [12]

3 Application on CNN

CNN is widely used in image segmentation, classification, detection and various other fields. Some of the examples are:

1. Face Recognition: CNN help in identifying unique features, focusing on each face despite of bad lighting, identifying all faces in picture [10].
2. Image classification: Image segmentation is the most challenging field in computer vision area which has been made easier by CNN approach [10].
3. Action Recognition: Various deformation of features in different patterns were achieved by CNN approach [11].
4. Human Pose Estimation and Speech Recognition: Wide variation in possible body poses were successfully predicted using CNN. It is also recently used in speech recognition as it is found that CNN works well for distant speech recognition, Noise robustness etc. [11].

4 Proposed Work to Detect Dust on Solar Panel

4.1 Dataset

We have taken data set of 30,000 images of dusty solar panels with labels of power loss keeping irradiance level as constant. We have collected data from our setup in solar lab from solar technology trainer kit as shown in Fig. 2, which is having a setup of halogen lamp, power supply and solar panel of 20 W. Solar panel is kept horizontal

Fig. 2 Experimental setup of solar technology trainer kit

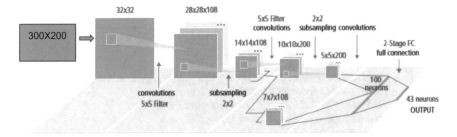

Fig. 3 Pictorial representation of our proposed model [10]

to halogen lamp, voltage and current generated were recorded through voltmeter and ammeter connected with the setup. Data was collected by keeping two identical solar panel one dusty panel and other reference panel with RGB camera. The effect of dust is calculated as the percentage loss in power output keeping irradiation as constant. The panel was exposed to different types of dust like red soil, sand, white cement, calcium carbonate etc. The dataset was collected for about a month with ample variation in dusting conditions from the experimental setup.

4.2 Simulation Environment

We have collected data from our experimental setup using RGB camera and collected total 30,000 solar panel images with power loss labels. Then we have converted the image file into steps and cropped the images to take the dusty portion in solar panel. We have also used data augmentation techniques. We have used CNN Lenet model. Figure 3 shows Lenet model that we have used in our work. We have RGB image to gray scale image and passed 300×200 image into two convolutional layers and pooling layers as shown in the figure and then finally passing it through fully connected layer and applying regression to compile the model and get the final output. The dataset was split randomly into training and validation data set. We have taken learning rate as 0.01, batch size as 65 and epochs as 10 and we have achieved mean squared error as 0.01 and accuracy as 80%. We have used convolutional layer, pooling layer, elu activation function, dropout layers and fully connected layers to build the model using keras environment. For optimization we have used adam optimizer with a learning rate as 0.01 and loss as mean squared error.

4.3 Sample Code

See Tables 1 and 2.

Table 1 The lenet model code that is used in this paper

```
def solar_model():
  model = Sequential()
  model.add(Conv2D(50,(5,5),input_shape=(28,28,3),activation='elu'))
  model.add(MaxPooling2D(pool_size=(2,2)))
  model.add(Conv2D(30,(3,3),activation='elu'))
  model.add(MaxPooling2D(pool_size=(2,2)))
  model.add(Conv2D(20,(2,2),activation='elu'))
  model.add(MaxPooling2D(pool_size=(2,2)))
  model.add(Dropout(0.5))
  model.add(Flatten())
  model.add(Dense(50,activation='elu'))
  model.add(Dense(25,activation='elu'))
  model.add(Dense(1,activation='elu'))
  model.compile(Adam(lr=0.01),loss='mse')
  return model
```

Table 2 Details of each layer obtained after executing our model algorithm

Layer	Output shape	Param
conv2d_19 (Conv2D)	(None, 24, 24, 50)	3800
max_pooling2d_19	(None, 12, 12, 50)	0
conv2d_20 (Conv2D)	(None, 10, 10, 30)	13,530
max_pooling2d_20	(None, 5, 5, 30)	0
conv2d_21 (Conv2D)	(None, 4, 4, 20)	2420
max_pooling2d_21	(None, 2, 2, 20)	0
dropout_7 (Dropout)	(None, 2, 2, 20)	0
flatten_7 (Flatten)	(None, 80)	0
dense_19 (Dense)	(None, 50)	4050
dense_20 (Dense)	(None, 25)	1275
dense_21 (Dense)	(None, 1)	26

Total params: 25,101
Trainable params: 25,101
Non-trainable params: 0

4.4 Outcomes

In this project we have used CNN approach to predict power loss in solar panel taking into consideration images of dusty solar panel along with their irradiance level. After carrying out CNN approach, the following graph in Fig. 4 is obtained which shows relationship between number of epochs/iterations with MSE (Mean squared error for training and validation data. From the graph we get conclusion that as the number

Fig. 4 Result of CNN Lenet based model showing relationship between no of epochs and MSE for training and validation data

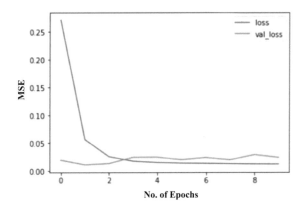

of epochs or iterations increases MSE i.e. termed as loss decreases and the loss for training data and validation data are nearer to each other we obtained mean squared error i.e. loss as 0.0122 and validation loss as 0.0241, which indicates that our error is less and our model is predicting 80% accurately.

We also tried to check with many random images the power loss that our model is predicting and compared it with power loss in our training dataset.

5 Conclusion

We have presented a CNN-based Lenet model approach for detection of dust on solar panel. We have taken RGB image of various dusty solar panel and predicted power loss due to dust deposition. We have used supervised learning method to train the model which avoids manual labelled localization. With this approach we have achieved mse as 0.0122.

Further more dataset should be collected in future to get better accuracy as well as some other models can be used to improve efficiency of the model.

References

1. D.S. Rajput, K. Sudhakar, Effect of dust on the performance of solar PV panel, in *International Conference on Global Scenario in Environment and Energy*, 4th–16th March 2013
2. S. Mehta, A.P. Azad, S.A. Chemmengath, V. Raykar, S. Kalyanaraman, DeepSolarEye: power loss prediction and weakly supervised soiling localization via fully convolutional networks for solar panels, in *2018 IEEE Winter Conference on Applications of Computer Vision (WACV)* (Lake Tahoe, NV, 2018), pp. 333–342
3. B.V. Chikate, Y.A. Sadawarte, The factors affecting the performance of solar cell. Int. J. Comput. Appl. (0975–8887), in *International Conference on Quality Up-gradation in Engineering, Science and Technology (ICQUEST2015)*

4. N.K. Memon, Autonomous vehicles for cleaning solar panels, in *Presented at International Renewable and Sustainable Energy Conference* (Marrakech, Morocco, 2016)
5. S. Pamir Aly, N. Barth, S. Ahzi, Novel dry cleaning machine for photovoltaic and solar panels, in *Presented at the 3rd International Renewable and Sustainable Energy Conference* (Marrakech, Morocco, 2015)
6. N.S. Najeeb, P. Kumar Soori, T. Ramesh Kumar, A low-cost and energy- efficient smart dust cleaning technique for solar panel system, in *Presented at 2018 International Conference on Smart Grid and Clean Energy Technologies* (2018)
7. M. Maghami, H. Hizam, C. Gomes, Impact of dust on solar energy generation based on actual performance, in *2014 IEEE International Conference Power & Energy(PECON)* (2014)
8. A. Hussain, A. Batra, R. Pachauri, An experimental study on effect of dust on power loss in solar photovoltaic module, Springer Open, Article number 9 (2017)
9. R. Yamashita, M. Nishio, R.K.G. Do, K. Togashi, Convolutional neural networks: an overview and application in radiology. **9**(4), 611–629 (2018)
10. S. Hijazi, R. Kumar, C. Rowen, IP Group, Cadence, Using convolutional neural networks for image recognition
11. A. Bhandare, M. Bhide, P. Gokhale, R. Chandavarkar, Applications of convolutional neural networks. (IJCSIT) Int. J. Comput. Sci. Inf. Technol. **7**(5), 2206–2215 (2016)
12. F. José Perales, J. Kittler, Articulated Motion and Deformable objects, in *9th International Conference, AMDO 2016*, Palma de Mallorca, Spain, 13–15 July 2016, p. 152

A Novel Method of Data Partitioning Using Genetic Algorithm Work Load Driven Approach Utilizing Machine Learning

Kiranjit Kaur and Vijay Laxmi

Abstract The data partition is a method which makes the processing of the database server's easy. It is like the clustering of the similar type of data files in an order so that the searching becomes easy. The data may be structured or unstructured. This paper focuses on the development of a unique data partition method which utilizes column integrated data. The proposed algorithm also utilizes Natural Computing Optimization inspired Genetic Algorithm (GA) for the improvisation of the partitioned data structure. The optimized set is cross validated utilizing Artificial Neural Network. This results into high values of evaluation parameters. The evaluation of the proposed algorithm is done using Precision, Recall and F-measure.

Keywords Data partition · Column family data · Genetic algorithm · Artificial neural network

1 Introduction

Data partition is made to simplify the search of the user query. If the data is well portioned then it will result in the high accuracy of data search. If the partition is not made well then a bunch of false data will fall under the search result. The search result is measured by precision and recall. The precision and recall are qualitative parameters of a search. High precision and recall indicate the good organization of the data.

Figure 1 demonstrates a well-partitioned data value. Every data is divided into four segments and the execution engine will find it smoothes to run the user query and the results will be good enough for the searched query. On the contrary to Figs. 1 and 2 presents a non organized data set.

K. Kaur (✉)
Guru Kashi University, Talwandi Sabo, Bathinda, India
e-mail: gill_kiran2004@yahoo.com

V. Laxmi
UCCA, Guru Kashi University, Talwandi Sabo, Bathinda, India
e-mail: deanca.gku@gmail.com

© Springer Nature Switzerland AG 2020
P. K. Mallick et al. (eds.), *Cognitive Computing in Human Cognition*,
Learning and Analytics in Intelligent Systems 17,
https://doi.org/10.1007/978-3-030-48118-6_5

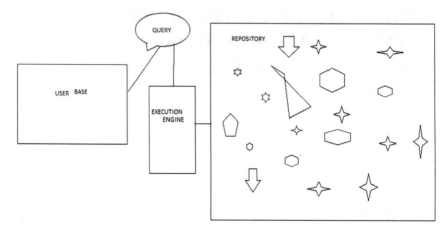

Fig. 1 A well-partitioned structure

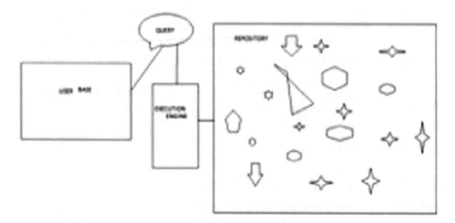

Fig. 2 Un-organized dataset architecture

The non organized dataset will always result in false data analysis and a false data pattern of retrieval against the user search. A column integrated data family is a family of data which is organized in columns. A data set can be viewed as a collection of data columns represented by equation number 1.

$$D = \{C1C2C3\dots Cn\} \tag{1}$$

The columns should be integrated in such a manner that each column co-related with the previous column. The integration of columns can be decided using equation number 2.

$$\text{Colvalue} = \sum \text{Coldata Raltion with Other Col Values} / \text{Total Number of Columns}$$
(2)

The similarity of the column values can be calculated using any similarity index as for cosine similarity, soft cosine similarity or Jaccard similarity. The similarity, item utilizes an arranged column family data. The rest of the paper is organized as follows.

The related work is discussed in Sect. 2 whereas the proposed model and methods are discussed in Sect. 3. The evaluation of the parameters and comparative analysis is done in Sect. 4 where as Sect. 5 concludes the paper.

2 Related Work

The basic of data partition is discussed in [1–7]. The cloud stores utilize the data partition for efficient transactions. Work load driven approach is discussed in [8]. The data partition mechanism presented handles the work load well and that inspires the proposed algorithm too. The scenarios of the vertical partitioning methods are discussed in [9–13]. Taking the work, one step further, the concepts of clustering is utilized in [14–16]. Efficient clustering mechanism plays a vital role in order to complete the transitions from data partitions [17–22] described the various methods of vertical partitioning and associated rule sets.

3 Proposed Model

The proposed work model is divided into two sections.

3.1 Identification of Unique Primary Key Values

The primary key (PK) is the unique identification key of any table. Algorithm 1 segregates all the primary key (PK) values from the data set.

```
Algorithm 1: Identify Primary_Key(Dataset)
Product_Positions[]
//An empty array is initialized to store the records
of the product
First_Key = Dataset.FirstValue
//Taking the first primary key value as reference
Productcounter = 0;
```

```
Foreachrowvalue in Dataset
//Taking each row of the dataset
If Dataset.Company.ProductId ==First_Key
//If the row value is equal to the reference record
Productcounter = productcounter + 1;
//Incrementing the product row number
Product_Position[Productcounter] = rowvalue;
End If
End For
RemovePK from datalist();
Repeat till all primary keys are not found.
End Algorithm
```

Algorithm 1 takes the first primary PK of the datatable and compares it with every row value and if the row value is same as that of the PK value then it stores its location. If no row value is left, it deletes the PK value from the list. Algorithm 2 is the proceeding of Algorithm 2. Algorithm 2 takes each location of the PK and stores it separately on cloud structure so that it becomes easy to fetch data when a query is made.

3.2 The Arrangement of Partition Values and Optimization

```
Algorithm 2: Store_Cloud_PK_References(Product_Position)
Foreachpos in Product_Position
//For every value in the product position value
Currentvalue=Product_Position(pos);
DesiredCol=ColUserInput;
//Taking the user requirement, which column he wants
to store
If Is Authenticated()
//Here if the user is authenticated at the cloud
server then the desired
Store_Structure[DesiredCol.values];
//column values will be stored at the cloud
EndIf
End Algorithm
```

It is not necessary that selected components by the user fall under appropriate necessity of the search query. Hence natural computing based Genetic Algorithm is applied to make sure that user only gets which is required. It will not only prevent

Table 1 Elements of the genetic algorithm

Algorithm type	Natural genetic
Population size	Total number of columns in the structure
Mutation value	Type Stoch
Cross over type	Linear/polynomial (based on column complexity)

the search engine from extra search time but will also reduce the complexity of the search. It optimizes the results [23] (Table 1).

```
Algorithm 3: Genetic Algorithm
[r,c] = size(Partioned_Data);
coldata = Partioned_Data
allassociation = [];
coldata1=unique(coldata);
for k = 1:numel(coldata1)
Lw=[];
cr=coldata1{k};
pos=0;
fori=1:numel(coldata)
ifstrcmp(cr,coldata{i})
pos=pos+1;
Lw(pos)=i;
end
end
allassociation(k)=numel(Lw);
Lpk=sum(Lw);
Ld=Lpk/no_of_partitions;
dLd(k)=sqrt((Lpk-Ld)^2/no_of_partitions);
end
allrank=allassociation+dLd;
[sortedrank,sortedpos]=sort(allrank,'descend');
sortedga = [];
counter=1;
fori=1:numel(sortedpos)
current=sortedpos(i);
currentversion=coldata1{current};
myallpos=[];
mycount=0;
for j = 1:numel(coldata)
ifstrcmp(currentversion,coldata{j})
mycount=mycount+1;
myallpos(mycount)=j;
end
```

```
end
forsn=1:numel(myallpos)
crp=myallpos(sn);
for n=1:c
sortedga{counter,n} = Partioned_Data{crp + 1,n};
end
counter=counter+1;
end
end
```

GA takes the partitioned data as input. It finds the unique available identities in
the list and for each unique element, the GA finds its alternative occurrence in the
entire sheet. Rank is calculated and sorted in descending order based on the total
occurrence count. The proposed work model is cross validated utilizing Artificial
Neural Networks (ANN) where output data of GA is used as input. ANN is a three
layer architecture namely input layer, hidden layer and output layer. The input layer
takes the raw data as input and a weight is generated by the hidden layer. The
data propagates in iterations and in each iteration the weight changes. Algorithm 4
demonstrate the structure of weight generation for hidden layer.

```
Algorithm 4: Weight Generator of Input Data
Generate_weight ()
Input: Data->d Output: Generated Weight- > W
Initialize W to Empty
W.size<-d. size
a= arbitrary constant
b= arbitrary constant
Foreach dl in d
W[dl] <-a*x+b //Linear weight mechanism
End For
Return W
```

Algorithm 4 is utilized to generate Weight for input data d. The weight generation mechanism uses two arbitrary constants "a" and "b". The input data element
d is first multiplied with the first arbitrary constant and then added to the second
arbitrary constant. The weight generation mechanism is always called when the
simulation iteration changes. ANN utilizes the generated weight to train and the
trained architecture is used to cross validate the input data.

```
Algorithm 5: Train Neural ()
   Input: Raw Data   <     −$D_R$   Output: Trained Data   $D_T$
Initialize $W_t$to empty
   //It is the weight which will be generated by
Algorithm 4
   Gradient$_{value}$ = Generate$_{Neural\ Graident}$
```

$\text{Total}_{\text{propagation iteration}} = 1000$

```
//A total of 100 simulation propagations is
considered
```

$G_{\text{satisfied}} = 0$

```
//Initially the gradient is not satisfied
```

$\text{Initialize } P_{\text{counter}} = 0;$

$\text{While Gradient}_{\text{value}}$ `is not satisfied &&`$P_{\text{counter}} < Total_{\text{propogationiteration}}$

```
//Either the       Pcounter is less than total provided iterations      or the
gradient is not satisfied, the loop will repeat itself
```

$W_{\text{new}} = \text{Generate}_{\text{weight}}()$

$$W_t = W_t + W_{\text{new}}$$

```
//Generating a new weight and adding it to new weight
```

$W_f \quad = \quad \sum_{k=0}^{n} W_{\frac{k}{n}}$ `//Taking the average weight for all the`

```
propagated iterations till now
```

$\text{If} W_f > \text{Gradient\_value}$

```
G_satisfied = 1
Else
P_counter++
D_r=W_f
End If
End While
Return D_r
End Algorithm
```

The Training algorithm of ANN takes the raw input data and updates its weight with each propagation. The propagation will continue until either the gradient is not satisfied or the propagation counter does not reaches to its maximum provided limit.

The trained neural structure is used to cross validate the partitioned data. The portioned data will act as the test data as supervised cross validation mechanism is considered here.

```
Algorithm 6: Simulate_Neural()
Input: Traine ⟦d_Neural⟧ _Data < -D_r
Output:AcceptedPartioned Value<- A_p
Test_Data = Traine ⟦d_Neural⟧ _Data
W_test=Generate_weight
If W_test≤D_r*80/100
//Taking 80% as the margin value
A_p [W_test ]=Accept Accept Partition Value Member
Else
A_p [W_test ]=Reject
Readjust Partition Value
```

```
End If
Return A_p
End Algorithm
```

The simulation part takes the trained neural structure and the raw input data as input. It is a supervised learning approach and hence the training data will act as the test data. The raw test data will be passed to the weight generation algorithm mentioned as Algorithm 4. If the generated test data does not vary more than 20% as that of the trained structure then the data will be accepted to be in correct partition else the partition will be readjusted.

4 Results

Based on the research algorithm, the following parameters are evaluated and analyzed.

4.1 Average Search Time (AST)

It is the total consumed time in order to fetch the results of the searched term. The proposed algorithm uses 1000 queries for the search.

$$AST = \sum Query\ Search \frac{Time}{Total\ Query\ Count} \tag{3}$$

Table 2 present the average search time in seconds for number of queries passed up to 1000 (Fig. 3).

Table 2 AST for queries

Total queries	AST in seconds
10	4
50	12
100	18
500	45
1000	96

Fig. 3 Pictorial
representation of AST
against passed queries

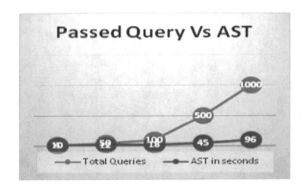

4.2 Precision and Recall

Precision and Recall are qualitative parameters. The high precision and recall value
represents the healthy arrangement of the dataset at the back end [24] (Fig. 4; Table 3).

$$\text{Precision} = \frac{\text{True}_{\text{Positive}}}{\text{True}_{\text{positive}} + \text{False}_{\text{Positive}}} \qquad (4)$$

$$\text{Recall} = \frac{\text{True}_{\text{Positive}}}{\text{True}_{\text{positive}} + \text{False}_{\text{Negative}}} \qquad (5)$$

Fig. 4 Pictorial
representation of precision
and recall

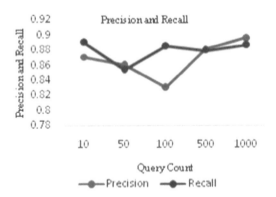

Table 3 Precision and recall
for same query set count

Query count	Precision	Recall
10	0.87	0.89
50	0.86	0.854
100	0.83	0.885
500	0.88	0.879
1000	0.896	0.886

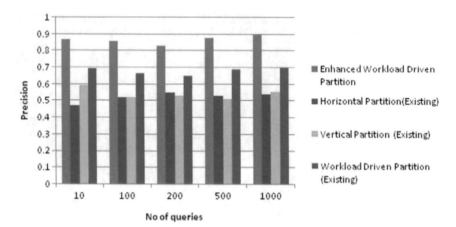

Fig. 5 Comparison of precision for different partition techniques

The high true positive rate represents good partition value. If the tp is high then it is clear that the data is kept in good partition cluster and there was no fuss in retrieving the data. It will also result into high precision and recall value. The proposed algorithm is passed around 1000 queries to the partition and the precision stood a good value of 0.88 on an average.

In paradox to tp, fp is the result of bad selection from the partition. High fp value represents that the elements were not placed in appropriate clusters and as a result the reverted results against the passed queries were also false. In the similar fashion, it will result into bad precision and recall value.

In proposed research work comparison with existing partition techniques on the basis of quality parameters are also performed. Figure 5 represents comparison of precision parameter for existing and proposed Enhanced partition.

Figure 6 represents comparison of Recall parameter for existing and proposed Enhanced partition.

5 Conclusion

This paper provides a fine solution for data partitioning and ways to remove the flaws of partitioning. The paper utilized column generated values and applies genetic algorithm in combination with Artificial Neural Network. ANN is used to cross validate the genetic algorithm verified data. A new fitness function is designed for genetic algorithm and the ANN uses feed forward architecture. A total of 1000 iterations are simulated and parameters like precision, recall and average search time is calculated. The proposed algorithm is also compared with several other algorithms which are listed. The proposed work has a lot of future possibilities. A combination

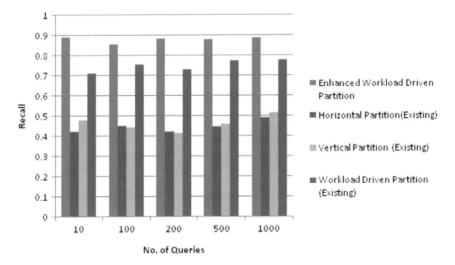

Fig. 6 Comparison of recall for different partition techniques

of Feed Forward Back Propagation Neural Network can be opted and tested for the same scenario. Other datasets can also be tried.

References

1. S. Ahirrao, R. Ingle, Scalable transactions in cloud data stores, in *IEEE 3rd International Advance Computing Conference (IACC)* (2013), pp 116–119
2. J. Baker, C. Bond, J. Corbett, J.J. Furman, A. Khorlin, J. Larson, J.-M. Leon, Y. Li, A. Lloyd, V. Yushprakh, Megastore: providing scalable, highly available storage for interactive services, in *CIDR*, vol. 11 (2011)
3. C. Curino, E. Jones, Y. Zhang, S. Madden: Schism: a workload-driven approach to database replication and partitioning, in *Proceedings of the VLDB Endowment*, vol 3 (2010), pp. 48–57
4. C. Curino, E.P.C. Jones, S. Madden, H. Balakrishnan: Workload-aware database monitoring and consolidation, in *Proceedings of the 2011 ACM SIGMOD International Conference on Management of data*, (2011), pp. 313–324
5. C. Curino, E.P.C. Jones, R.A. Popa, N. Malviya, E. Wu, S. Madden, H. Balakrishnan, N. Zeldovich, Relational cloud: a database-as-a-service for the cloud (2011)
6. S. Das, S. Agarwal, D. Agrawal, A.E. Abbadi, C. Bunch, N. Chohan, C. Krintz, J. Chohan, J. Kupferman, P. Lakhina: Elastras: an elastic, scalable, and self managing transactional database for the cloud, in *Technical Report 2010–04*, CS, UCSB (2010)
7. S. Das, D. Agrawal, A. El Abbadi (2009) Elastras: an elastic transactional data store in the cloud, in *USENIX HotCloud*, 2 (2009)
8. M. Liroz-Gistau, R. Akbarinia, E. Pacitti, F. Porto, P. Valduriez, Dynamic workload-based partitioning for large-scale databases, in *Database and Expert Systems Applications* (Springer. Berlin, 2012), pp 183–190
9. P.A. Bernstein, I. Cseri, N. Dani, N. Ellis, A. Kalhan, G. Kakivaya, D.B. Lomet, R. Manne, L. Novik, T. Talius, Adapting microsoftsql server for cloud computing, in *Data Engineering (ICDE). IEEE 27th International Conference* (2011), pp 1255–1263

10. S. Agrawal, V. Narasayya, B. Yang, Integrating vertical and horizontal partitioning into auto-mated physical database design, in *Proceedings of the 2004 ACM SIGMO International Conference on Management of Data* (2004), pp. 359–370
11. C. Sharma, J. Muthuraj, R. Varadarajan, S. Navathe, An objective function for vertically parti-tioning relations in distributed databases and its analysis, in *Distributed and Parallel Databases*, vol. 2(2) (1994), pp 183–207
12. W.W. Chu, I.T. Leong, A transaction-based approach to vertical partitioning for relational database systems. IEEE Trans. Softw. Eng. **19**(8), 804–812 (1993)
13. D.W. Cornell, P.S. Yu, An effective approach to vertical partitioning for physical design of relational databases. IEEE Trans. Software Eng. **16**(2), 248–258 (1990)
14. J.A. Hoffer, D.G. Severance, The use of cluster analysis in physical data base design, in *Proceedings of the 1st International Conference on Very Large Data Bases* (ACM, 1975), pp. 69–86
15. Huang, Y.-F., Lai, C-J., Integrating frequent pattern clustering and branch-and-bound approaches for data partitioning. Inf. Sci. 288–301 (2016)
16. A. Jindal, J. Dittrich, September) Relax and let the database do the partitioning online, *International Workshop on Business Intelligence for the Real-Time Enterprise* (Springer, Berlin, 2011), pp. 65–80
17. J. Kamal, M. Murshed, R. Buyya, Workload aware incremental repartitioning of shared-nothing distributed databases for scalable OLTP applications. Fut. Gener. Comput. Syst. 421–435 (2016)
18. S. Navathe, M. Ra (1989) Vertical partitioning for database design: a graphical algorithm. ACM SIGMOD Record **18**(2), 440–450
19. S. Navathe, S. Ceri, G. Wiederhold, J. Dou, Vertical partitioning algorithms for database design. ACM Trans. Database Syst. (TODS) **9**(4), 680–710 (1984)
20. S. Papadomanolakis, A. Anastassia, Autopart: Automating schema design for large scientific databases using data partitioning, in *Scientific and Statistical Database Management. 16th International Conference on IEEE Proceedings* (2004) pp. 383–392
21. S. Phansalkar, A. Dani, Transaction aware vertical partitioning of database (TAVPD) for respon-sive OLTP applications in cloud data stores. J. Theoret. Appl. Inf. Technol. **59**(1), 73–81 (2016)
22. J.H. Son, M.H. Kim, An adaptable vertical partitioning method in distributed systems. J. Syst. Softw. **73**(3), 551–561 (2004)
23. D.E. Goldberg, J.H. Holland, Genetic algorithms and machine learning. Mach. Learn. **3**(2), 95–99 (1988)
24. M. Sharma, G. Singh, R. Singh, Stark assessment of lifestyle based human disorders using data mining based learning techniques. IRBM **38**, 305–324 (2017)

Virtual Dermoscopy Using Deep Learning Approach

Debabrata Swain, Sanket Bijawe, Prasanna Akolkar, Mihir Mahajani,
Aditya Shinde, and Pradunya Maladhari

Abstract Dermoscopy is one of the most irregular and challenging areas to diagnose as it is very complex. In the sphere of dermatology, numerous numbers of times, thorough examinations are required to be carried out to resolve upon the skin ailment the patient may be facing. Different practitioners may take a different amount of time to detect the skin disease. So, a system is required that can efficiently and accurately diagnose the skin conditions without any such restrictions. This paper presents an automated dermatological diagnostic system using a deep learning approach. Dermatology is the branch of medicine which deals with the identification and treatment of skin diseases. The presented system is a machine interference in contradiction to the traditional medical personnel-based belief of dermatological diagnosis. The entire system works on the two mutually dependent steps. The first is preprocessing of image of that part of skin that is infected and the second step is used to recognize the disease. The system uses convolutional neural networks and feedforward backpropagation for the identification of skin disease. The system gives an accuracy of 93.063% while testing on a total of 180 image samples for six disease classes.

Keywords Automated · Backpropagation · Classification · Dermatology ·
Diagnostic · Feedforward

D. Swain (✉) · S. Bijawe · P. Akolkar · M. Mahajani · A. Shinde · P. Maladhari
IT Department, Vishwakarma Institute of Technology, Pune, India
e-mail: debabrata.swain@vit.edu

S. Bijawe
e-mail: sanket.bijawe18@vit.edu

P. Akolkar
e-mail: prasanna.akolkar18@vit.edu

M. Mahajani
e-mail: mihir.mahajani18@vit.edu

A. Shinde
e-mail: aditya.shinde18@vit.edu

P. Maladhari
e-mail: pradunya.maladhari18@vit.edu

© Springer Nature Switzerland AG 2020
P. K. Mallick et al. (eds.), *Cognitive Computing in Human Cognition*,
Learning and Analytics in Intelligent Systems 17,
https://doi.org/10.1007/978-3-030-48118-6_6

1 Introduction

Some of the most widespread diseases across the globe are dermatological diseases. In spite of being so common, their detection is very confusing and needs thorough expertise in that field. It is estimated that more than 9500 people in the United States are diagnosed with skin cancer every day. There is irregular schooling in dermatology at the college level, which shows that the trainees should consider reevaluating their existing skills. Right now, about 90% of skin diseases are being managed only by using primary care. These imply that if we take care at an early stage, most of the skin disease dilemmas could be resolved. Skin diseases can have a significant impact on the quality of life of patients. The number of skin diseases are increasing constantly and results obtained are dependent on the initial diagnosis. Some people cannot afford to go to a dermatologist for their skin problems. In the countries that are not technologically forward, there have been numerous attempts to implement traditional medicine. But the attempts have met with some problems such as the shortage of medical expertise and a very high cost of medical tools. Skin diseases are a result of environmental factors as well as various other causes. The important tools that are necessary for the timely detection of such diseases are not very easily available to a large percentage of the population worldwide, which is one of the major reasons for a very high mortality rate in these cases. This paper provides a method to detect some of the types of skin diseases. This project aims to collect skin disease image data, preprocesses that data and creates an intelligent prediction system for better identification of skin diseases. The image of the affected skin part is provided by the user as an input to the system, which then performs processing on it by extracting features using the Convolutional Neural Networks algorithm. To diagnose skin diseases, it uses a SoftMax image classifier. The proposed system will thus be very useful for those places which have a very limited access to medical facilities. It can also be a very handy tool for the doctors to verify their diagnosis in case of those diseases which cannot be detected accurately in their initial stages, like melanoma. Thus, this paper proposes skin disease identification and classification method based on Convolutional Neural Network. In our system, we have worked on classifying 6 different skin ailments listed as Acne and Rosacea, Bullous Disease, Cellulitis Impetigo and other Bacterial Infections, Eczema, Melanoma Skin Cancer Nevi, and Nail Fungus. We have made use of Image Processing and Artificial Neural Networks to classify the different types of skin diseases. To convert an input image to an output label we are using a convolutional neural network. It is particularly designed for image data processing. In this paper, we will be discussing the architecture, methodology and pre-processing algorithms used in our system.

2 Background

2.1 A Brief Overview on Related Works

Rathod et al. [1] proposed a method for detecting various kinds of skin diseases. CNN algorithm was used for feature extraction along with an image classifier as SoftMax to diagnose diseases. Initially, an accuracy of about 70% was achieved. Their accuracy was then increased by increasing the training data. Further accuracy of 90% was achieved on 6 diseases. Yasir et al. [2] used computer vision along with image processing and fed the data to artificial neural networks. This system, when tested on a dataset of 775 images of 9 different disease types, it gave an accuracy of 90%. Asghar et al. [3] proposed a rule-based expert system was developed. This system used forward chaining along with a depth-first search to detect diseases. The paper by Amarathunga [4] gave an approach which could diagnose the skin disease as well as give medical treatment or advice quickly. That system used image processing as well as data mining. To enhance the image, different preprocessing techniques were used. Finally, the use of data mining techniques for suggestions of treatment and advice provided them with accuracies of 85%, 95% and 85% for Eczema, Impetigo, and Melanoma respectively. In another paper published by Liao et al. [5] a universal skin disease detection system was constructed using CNN. They got datasets from Dermnet and OLE. They had 73.1% Top-1 accuracy and 91.0% Top-5 accuracy on Dermnet and 31.1% Top-1 and 69.5% Top-5 accuracy on OLE. Lopez et al. [6] used VGGNet convolutional neural network architecture along with the transfer learning paradigm. They achieved a sensitivity value of 78.66% as well as 79.74% precision while working on ISIC Archive dataset.

2.2 Database Description

Tons of data surrounds us today, generated from various instances by various sources. Each data generated belongs to different and heterogeneous formats, each of several different domains. This collected data can prove to be immensely valuable if used properly. There are various approaches to obtain live data for research and development. The skin disease database used is scraped from "Dermnet"—a web portal that contains a total of 23,000 images which is further classified in 46 diseases. Our scraped database contains 6 diseases which are a total of 1600 images. Figure 1 flow diagram describes our approach towards dataset scraping. All the labels of the scrapped disease are listed as follows:

1. Acne-and-Rosacea-Photos
2. Bullous-Disease-Photos
3. Cellulitis-Impetigo-and-other-Bacterial-Infections
4. Eczema-Photos

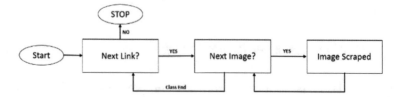

Fig. 1 Flow diagram for dataset scraping

5. Melanoma-Skin-Cancer-Nevi-and-Moles
6. Nail-Fungus-and-other-Nail-Disease.

3 The Proposed Methodology and Implementation

3.1 Brief Overview of System

The proposed methodology, depicted in Fig. 2, is based on two primary parts, Image pre-processing unit and a classifier unit. The image processor unit will augment the sample/image by reshaping it and then the image will be divided into segments. Then the image will be sent to the convolutional classifier for feature extraction and further classification.

- **Image Processing Unit**: This unit focuses on the affected part by converting the image into the RGB form. Feature extraction is a major step in the classification problem as it is the core of the image classification problem. So, both the training and testing images are resized into proper format before sending it to the classifier unit. Furthermore, all the training images are passed through the rotational and

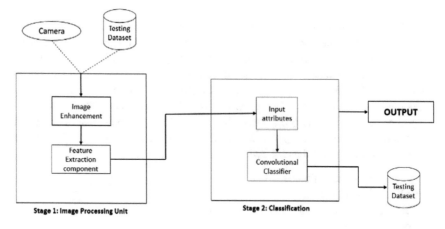

Fig. 2 Architecture of proposed model

positional variant i.e. they are shifted vertically or horizontally, rotated clockwise or anti-clockwise by 10% or scaled in/out, thus sustaining the process of efficiently training the model [7].

- **Classification Unit**: This unit classifies the images into pre-defined classes using a convolutional neural network algorithm and SoftMax classifier for multi-class classification.

3.2 ConvNets Architecture

A **Convolutional Neural Network (CNN or ConvNet)** is a Deep Learning algorithm that is capable of taking an image input and assigning a value (learnable biases and weights) to several features in the images which can initiate the differentiation from one another. The required pre-processing in a ConvNet is enormously low in contrast to other classification algorithms. Whereas in primary methods, features are hand-engineered and with adequate training, ConvNets learns these characteristics by itself.

ConvNet is preferred over a feed-forward neural network as it is able to **strongly catch the Temporal and Spatial dependencies** in an image by applying the appropriate filters. Due to the decline in the total number of parameters connected and reusing the ability of weights, the architecture presents an apt fixture of the image dataset. In simpler words, for a better knowledge of refinement of the image, the network can be trained. Figure 3 represents the ConvNet architecture designed by us.

The Four main layers used in ConvNets architecture is listed as below:

- **Convolution**: The main advantage of the Convolutional technique in ConvNets is for differentiating appropriate features from the image that can act as an input for the initial layer. The spatial interrelation of the pixels is provided by the convolution [8]. Figure 4 shows the convolution of a matrix using (3, 3) as its

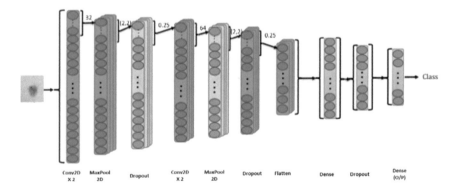

Fig. 3 Architecture of CNN

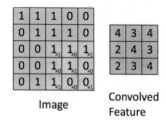

Image   Convolved
                     Feature

Fig. 4 Convolution of a matrix

kernel size. Equation (1) gives the statistical representation for convolution of a matrix.

$$I\_new(x, y) = \sum_{j=-1}^{1} \sum_{i=-1}^{1} a_{ij} I\_old(x - i, y - j) \qquad (1)$$

- **Activation Function**: Rectified Linear Unit ReLU is used in the model to increase non-linearity in the ConvNet and it acts on a fundamental step. In other words, it is an advancement that is implemented per pixel and outmodes the non-positive values of every pixel, hence mapping the feature by zero. It is an uniform approximation and is used for multiclass classification in our model [9]. Figure 5 shows the graph of the ReLU function. Equation (2) represents the ReLU function.

$$R(x) = \max(0, y)n \qquad (2)$$

- **Pooling**: Spatial Pooling, also known as subsampling or downsampling benefits in the reduction of the dimensions of each feature mapping but while doing so, it retains the most significant information of the map. MaxPool helps in finding the maximum value of the pixels present in the provided kernel size, thus enhancing the features available in the sample [10]. After pooling is done, our three-dimensional feature map gets converted to the single-dimensional column

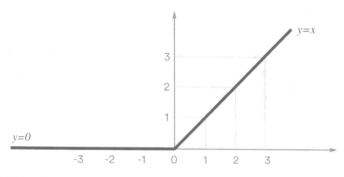

Fig. 5 ReLU function

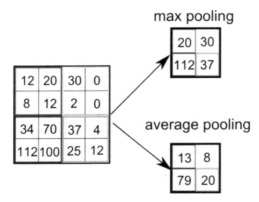

Fig. 6 MaxPool and average pooling of a matrix

vector. Figure 6 shows the max pooling and average pooling of a matrix using (2, 2) as kernel size.

- **Flatten**: Flattening is an essential step in ConvNets. Flattening is done as soon as the pooled featured map is obtained. It involves modification of the complete pooled feature mapped matrix into a single column. Furthermore, it is fed to the artificial neural network for processing. Figure 7 is a matrix representation of the technique used in flattening.

- **Classification (Fully Connected Layer)**: After flattening is done, the flattened feature mapped matrix is sent through a fully connected neural network. This step is a combination of the fully connected layer and the output layer. The fully connected layer is alike the hidden layer in artificial neural network's but in this case, it's fully connected. We get the predicted classes at the output layer. The error of prediction is calculated on the basis of the passed information. Then backpropagation of error is done through the system in order to improve the weights and the biases. Figure 8 shows the connections between fully connected layer/dense layer.

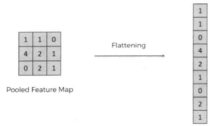

Fig. 7 Flatting of a matrix

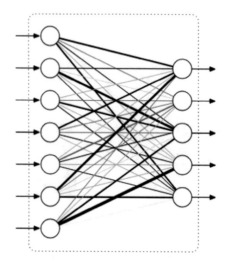

Fig. 8 Fully connected layer

4 Results and Discussions

This research comprises of six diagnosis classes as mentioned in the database description. As transcribed earlier in the section of 'A brief overview on related works', some researchers stated several techniques for diagnosing skin disease. The accuracies reported by them vary between 50 and 90%.

The database used in our whole system consists of 1600 samples of which around 1400 are complete samples used for training and validating the ConvNet model and 180 images are samples with missing attributes used for gaining testing accuracies. Of 1400 samples, 85% of the database was used for training the ConvNet and rest of the database (15%) was used for validating the proposed system.

The confusion matrix in the heatmap form demonstrates the classification results of the system. In this matrix, each cell contains the analogous accuracies of exemplars classified for the corresponding combination of desired and actual network outputs [11]. Figure 9 gives the confusion matrix displaying the classification results of this network. The y-axis and x-axis denote the labels of actual and predicted classes. The vertical scale in Fig. 9 denotes the color to the accuracy pattern in the heatmap.

Figure 10 shows the AUC curve i.e. accuracy curve and loss function. It shows how accuracy was eventually increased on training epochs. As it displays, we've successfully achieved an accuracy of 93.063% on a testing dataset containing a total of 180 images of different classes. Also, the loss function plot shows the avoiding of overfitting throughout the processing (training). The AUC score obtained from training is 0.841664.

Table 1 shows the report of classification used for assuring the classification quality. The report presents the principal classification metrics support, F1-score, recall, and precision on a per-class basis.

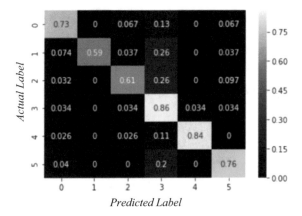

Fig. 9 Confusion matrix

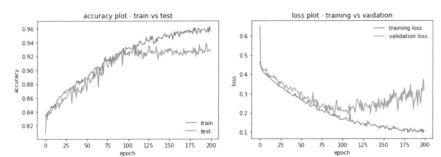

Fig. 10 Accuracy and loss plot

Table 1 Classification report

	Precision	Recall	F1-score	Support
0	0.793	0.767	0.780	30
1	1.000	0.615	0.762	26
2	0.800	0.625	0.702	32
3	0.481	0.862	0.617	29
4	0.970	0.842	0.901	38
5	0.760	0.760	0.760	25
Accuracy			0.750	180
Macro avg	0.801	0.745	0.754	180
Weighted avg	0.807	0.750	0.760	180

- Precision—Ability of a classifier not to label an occurrence positive that is negative.
- Recall—Ability of a classifier to find all positive occurrence.
- F1 score—Weighted harmonic mean of precision and recalls.

The overall F1-score of our model is 0.750, tested on 180 images.

5 Limitations

- No predictions can be made in absence of any skin disease due to the unavailability of appropriate dataset for normal skin.
- No application is available to provide the user with an interface to interact with the model.
- The accuracy of the model can be increased by importing the datasets directly from the hospitals to get more authentic information.

6 Future Scope

- Creating an android application which will be able to make real time predictions.
- Suggesting specialized doctors to the user based on the predictions made by the application.
- Dynamic dataset updating by the doctors so as increase the accuracy of the system.

7 Conclusion

In this paper, we have learned about how the image classification is done using ConvNets. Thus, skin diseases can be diagnosed as well as classified using this technique. Currently, skin diseases have been classified into 6 different types with around 230 images in each class. The experimental results gained a classification accuracy of 93.33% on the testing set. The use of large datasets and advanced computational techniques will help in increasing the accuracy of the system thereby allowing it to suit the results of a dermatologist. This will in turn help in uplifting the quality standards in the field of medicine and research.

Acknowledgements We would like to thank our guide Mr. Debabrath Swain for helping us with his invaluable experience and also his constant motivation has helped us complete our project successfully.

References

1. J. Rathod et al., Diagnosis of skin diseases using convolutional neural networks, in *Proceedings of the 2nd International Conference on Electronics, Communication and Aerospace Technology (ICECA 2018)*. IEEE Conference Record # 42487; IEEE Xplore. ISBN: 978-1-5386-0965-1
2. R. Yasir, M. Rahman, N. Ahmed,*Dermatological disease detection using image processing and artificial neural network*, January 2015
3. M. Asghar, M. Asghar, S. Saqib, B. Ahmad, Diagnosis of skin diseases using online expert system. Int. J. Comput. Sci. Inf. Secur. **9**(6), 323–325 (2011)
4. A. Amarathunga et al., Expert system for diagnosis of skin diseases. Int. J. Sci. Technol. **4**(1) (2015)
5. H. Liao, *A Deep Learning Approach to Universal Skin Disease Classification*, (Department of Computer Science, University of Rochester) (2015)
6. A.R. Lopez, X. Giro-i-Nieto, Skin lesion classification from dermoscopic images using deep learning techniques, in *Proceedings of the IASTED International Conference Biomedical Engineering (BioMed 2017)*, Innsbruck, Austria, 20–21 Feb 2017
7. D. Swain, S.K. Pani, D. Swain, An efficient system for the prediction of coronary artery disease using dense neural network with hyper parameter tuning. Int. J. Innov. Technol. Explor. Eng. (IJITEE) **8**(6S) (2019). ISSN: 2278-3075
8. D. Swain, S.K. Pani, D. Swain, Diagnosis of coronary artery disease using 1-D convolutional neural network. Int. J. Recent Technol. Eng. (IJRTE) **8**(2) (2019). ISSN: 2277-3878
9. Identification of erythemato-squamous skin diseases using extreme learning machine and artificial neural network. ICTACT J. Soft Comput. **04**(01) (2013)
10. Use of neural network-based deep learning techniques for the diagnostics of skin diseases. Biomed. Eng. (2019)
11. Diseases by combining deep neural network and human knowledge, in *The 2nd International Workshop on Semantics-Powered Data Analytics*, 23 July 2018

Evaluating Robustness for Intensity Based Image Registration Measures Using Mutual Information and Normalized Mutual Information

Navdeep Kanwal, Shweta Jain, and Paramjeet Kaur

Abstract Numerous Applications envisage the need of image registration. The term image registration is the process of alignment of multimodal, multitemporal or multi-view images onto a single co-ordinate system by applying certain transformations e.g. rotation, translation, scaling etc. This paper uses Information Measures such as Mutual Information (MI), and Normalized Mutual Information (NMI) to obtain the aligned image and then evaluate their robustness. Various Parameters such as Pearsons correlation coefficient (PCC), Peek Signal to Noise Ratio (PSNR), have been used to quantify the image registration results obtained through MI and NMI. The methodology for registration has been validated using above parameters and elapsed time is also evaluated for rigid registration of natural Images.

Keywords Image registration · Mutual information · PCC · PSNR

1 Introduction

Data set of images acquired through sampling at various instances of time from different perspectives lie on different coordinate systems. Capturing the information from these different modalities foresee the need of image registration for alignment of images onto a single coordinate system. It involves finding transformations that relates information conveyed in floating or sensed image to be registered on reference or fixed image [1, 2]. Because of the variety of images, application areas, registration methods a single registration method may not be suitable for all types of images [3].

N. Kanwal (✉) · S. Jain
Department of Computer Engineering, Punjabi University Patiala, Punjab, India
e-mail: navdeepkanwal@pbi.ac.in

S. Jain
e-mail: shwetajn1991@gmail.com

P. Kaur
Department of Computer Science and Engineering, Baba Farid Institute of Engineering and Technology Bathinda, Punjab, India

© Springer Nature Switzerland AG 2020
P. K. Mallick et al. (eds.), *Cognitive Computing in Human Cognition*,
Learning and Analytics in Intelligent Systems 17,
https://doi.org/10.1007/978-3-030-48118-6_7

Nevertheless all registration methods include feature extraction, feature matching, transformation selection and image resampling [4, 5]. Different researchers like [6, 7] have proposed and discussed various methods of image registration. This paper focuses on Image Registration using Maximization of Mutual Information (MI) between two images as base for registration. Mathematical sciences have an important role of Mutual Information [8]. Results have been quantified using various parameters such as Pearsons Correlation Coefficient (PCC), peek signal to noise ratio (PSNR) and elapsed time (ET). The paper is organized as different sections. Section 2 introduces similarity metrics MI and NMI and various parameters used to validate results. Section 3 presents the methodology adopted and presents comparison between the similarity metrics. Section 4 presents the results computed on two different images and dataset of six image pairs [9] and validated with different parameters. Finally in subsequent section conclusion and future scope has been discussed.

2 Registration

2.1 Mutual Information

Mutual information (MI) is a statistical measure that finds its roots in information theory [10, 11]. It is a measure of how much information one random variable contains about another [12, 13]. The information to be registered in both the sensed and referenced image is same but the amount of information which is to be registered depends on the registration transformations. The information common amongst both the images is measured through joint entropy and this process is known as Mutual Information [3, 14]. The mutual information of two images A and B has been defined in Eq. (1) [12]:

$$I(A, B) = H(A) + H(B) - H(A, B) \tag{1}$$

where H (A) and H (B) are the Shannon entropies defined in Eq. (2) and Eq. (3), and H (A, B) is the Shannon entropy of the joint distribution of A and B [12, 15].

$$H(A) = - \sum_{a \in A} P_A log P_A(a) \tag{2}$$

$$H(B) = - \sum_{b \in B} P_B log P_B(b) \tag{3}$$

$$H(A, B) = - \sum_{a \in A} P_A \sum_{b \in B} P_B log P_A P_B(a, b) \tag{4}$$

where a and b are the intensity values in image A and B, P_A and P_B the marginal probability distribution of A and B, respectively, and $P_A P_B$ is the joint probability of A and B in Eq. (4). Corresponding intensities a in image A and b in B are related through geometric transformation [16].

Mutual Information metric, expressed as I(A, B) in Eq. (1) has been used in the work. It is defined as a measure of how well one image explains the other and is maximized at the optimal alignment of images. MI of images A and B is given in Eq. (5):

$$MI(A, B) = \sum_a \sum_b P_{AB}(a, b) log \frac{P_{AB}(a, b)}{P_A(a) P_B(b)} \tag{5}$$

A variant of MI i.e. Normalized Mutual Information (NMI) [17, 18] has been addressed as it is more overlap independent. In certain cases MI does not behave well For example, changes in information of low intensity region can disproportionately change MI. Hence, to overcome this problem of MI normalizations of joint entropy i.e. NMI is discussed. NMI may be derived as in Eq. (6).

$$NMI(A, B) = \frac{H(A) + H(B)}{H(A, B)} \tag{6}$$

2.2 Measures for Registration Quality

Pearsons Correlation Coefficient (PCC): To determine the strength of relationship between two variables PCC may be computed [19]. Its values lies between -1.00 and 1.00. Negative value indicates negative correlation i.e. with increase in value of one variable other decreases and positive value indicates positive correlation.

$$P_{xy}(x, y) = \frac{cov(x, y)}{\sigma_x \sigma_y} \tag{7}$$

where, σ_x and σ_y are the standard deviations, and cov is the covariance of X and Y in Eq. (7).

Peak Signal to Noise Ratio (PSNR): PSNR is the ratio between maximum power of a signal and the power of the corrupting noise that affects the fidelity of its representation as defined in Eq. (8). It is usually expressed in logarithmic decibel scale. It can range between 1 to infinity. Higher the value higher is quality of the image.

$$PSNR = 10 log_{10} \frac{(Max)^2}{MSE} \tag{8}$$

3 Methodology

Intensity based registration involves calculating the registration transformation by iteratively optimizing some similarity measure calculated directly from the pixel values in the images rather than from geometrical structures such as points or structures derived from images [20, 21].

The reference and sensed images are loaded and pre-processed for ensuring the size of images required for accurate registration. Initial data points are selected in sensed image space and passed to the optimizer such that it maximizes image similarity measure (SM) i.e. mutual information by changing transformation parameters i.e. rotation and scaling. It gives registered image containing maximum of mutual information between reference and sensed image. Figure 1 elaborates the process.

The starting point is required for an optimization that is the algorithm takes a series of guesses to choose starting position for performing registration. The starting position has to be close for the algorithm to perform accurate transformations. The algorithm then computes similarity function such as MI, joint entropy, correlation coefficient etc. relating to how well the two images can be registered. Some functions increase and others decrease. The registration algorithm proceeds by recalculating the similarity functions and the optimum is reached by seeking the transformation at which the maximum function is obtained. The starting point, type of transformations and the optimization process has a large influence on the results and robustness of the registration method.

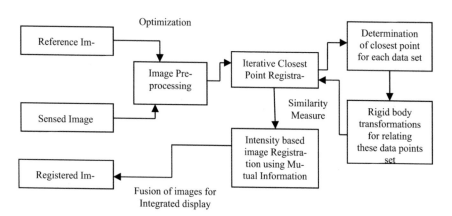

Fig. 1 Methodology for image registration

3.1 Robustness

Robustness of image registration depends upon different factors [22]. In this paper Similarity Measures (SM) robustness depends on the type of image and the transformations used. SM robustness for spatial transformations such as rotation and scaling is computed.

Rotation For rotation, an evaluation range of $[0, +5]$ and $[-5,0]$ degrees is considered. The selected SMs (MI and NMI) behave symmetrically around the peek point.

$$T(v) = Rv + t \tag{9}$$

where R is rotation matrix, t is a translation vector, $T(v)$ is the transformed vector. **Scaling** For scaling, both stretching and shrinking show different results on images. So down scaling i.e. when scaling factor is smaller than 1 is performed at both 0.5 and 0.75 and Up scaling i.e. when scaling factor is larger than 1 is performed at 1.25 value.

4 Results and Discussion

Results have been obtained by implementing the methodology described in Fig. 1 over a set of images. This registration methodology can work on a large variety of image types, and real scenes captured under different lightening conditions. Six image pairs captured from a data set of 22 image pairs has been used to evaluate the results. The dataset includes image pairs taken of indoor and outdoor scenes, in natural and manmade environments, at different times of day, during different seasons of the year, and using different imaging modalities. It includes image pairs with low overlap (e.g. 2%), substantial differences in orientation (90 degrees), and large changes in scale (up to a factor of 6.4).

4.1 Results of Mutual Information

Method described above is applied to different images with different scaling (0.5, 0.75, 1.25) and at different angles of rotations (ranging from -20 to $+40$). The resulting values for different metrics are tabulated.

Method is applied on images and results obtained for two of images are tabulated in Table 1. With the Scaling at 0.5 when the Image is rotated at angle 15 clockwise and maximum value of MI i.e. 0.9502 and 0.8429 is obtained for image 1 and image 2 respectively and it decreases with increase in angle. Similarly when at same scaling if image is rotated anti-clockwise MI at -15 is 0.9464 for image 1 and 0.8415 for

Table 1 Results of MI and other metrics for image 1 and image 2

Scale	Angle	Image 1				Image 2			
		MI	PCC	PSNR	ET(sec)	MI	PCC	PSNR	ET(sec)
0.5	−20	0.9207	0.9240	10.92	16.26	0.8359	0.0535	9.73	14.13
	−15	0.9464	0.8776	11.11	14.58	0.8415	0.0778	9.75	14.59
	15	0.9502	0.8996	11.97	16.19	0.8429	0.0456	11.13	16.79
	20	0.9464	0.9487	11.32	13.40	0.8391	0.0451	10.19	13.56
	25	0.9458	0.9445	10.90	15.96	0.8374	0.0519	9.65	13.80
0.75	−20	0.9351	0.8786	11.24	18.79	0.8443	0.3805	9.87	16.64
	−15	0.9573	0.8306	11.17	15.98	0.8553	0.4193	10.81	13.37
	5	0.9469	0.8292	11.78	15.46	0.8545	0.4485	11.47	16.29
	15	0.9745	0.8409	11.44	14.72	0.8577	0.4111	10.42	16.32
	20	0.9621	0.9814	10.61	15.48	0.8490	0.4016	10.21	15.73
	25	0.9572	0.9311	10.07	18.56	0.3617	0.3729	9.18	18.30
1.25	−5	0.9545	0.5156	12.76	19.93	0.8496	0.4941	12.92	20.97
	5	0.9060	0.4966	12.86	18.81	0.8472	0.5057	12.98	18.16
	10	0.2574	0.6283	6.44	18.60	0.8643	0.3850	10.21	17.70

image 2. It decreases as image is rotated anti-clockwise. PSNR is maximum at angle 15 that is 11.97 and 11.13 respectively whereas value of PCC ranges from 0.8776 to 0.9487 for image 1 and 0.0451 to 0.0778 for image 2. Elapsed time i.e. time taken to execute the procedure is also computed and presented in seconds. With the increase in scaling to 0.75 MI also increases and MI is maximum at angle 15 for both the images i.e. 0.9745 and 0.8545 respectively. MI decreases with increase in angle. Similarly results are also computed at scaling of 1.25 and rotations at different angles.

4.2 Summary of Mutual Information

Mutual Information decreases with increase in angle clockwise and when angle is increased anti-clockwise MI decreases. With increase in MI all other parameters improves and degrades when MI decreases. When there is increase in the Scaling we rotated the images at comparatively smaller angles as with increased scaling and big angle rotations MI almost declines giving poor quality image.

4.3 Results of Normalized Mutual Information

Method described in Fig. 1 is applied to different images with different scaling (0.5, 0.75, 1.25) and at different angles of rotations (ranging from −20 to +40). The resulting values for different metrics are tabulated below.

Results obtained for two images are tabulated in Table 2. With the Scaling at 0.5 when the Image is rotated at angle 15 clockwise and maximum value of NMI i.e.

Table 2 Results of NMI and other metrics for image 1 and image 2

Scale	Angle	Image 1				Image 2			
		NMI	PCC	PSNR	ET(sec)	NMI	PCC	PSNR	ET(sec)
0.5	−20	1.075	0.9264	11.06	14.81	1.0816	0.4033	10.43	15.31
	−15	1.077	0.8714	11.23	16.48	1.0810	0.4335	11.28	13.17
	15	1.0804	0.9091	11.24	17.62	1.0814	0.4488	11.28	14.12
	20	1.0796	0.9288	11.21	14.29	1.0813	0.4084	10.39	15.29
	25	1.0795	0.9632	10.63	17.27	1.0818	0.3963	10.05	13.52
0.75	−20	1.0778	0.8774	11.50	15.31	1.0812	0.4534	11.66	14.14
	−15	1.0784	0.7740	11.74	18.73	1.0831	0.4352	11.11	16.73
	5	1.080	0.7731	11.74	16.13	1.0815	0.4970	12.58	14.81
	15	1.0824	0.8652	11.35	14.98	1.0815	0.4813	12.36	16.86
	20	1.0850	0.9476	9.94	16.47	1.0819	0.4233	10.78	17.12
	25	1.0855	0.9197	9.80	15.77	1.0825	0.3817	9.54	15.89
1.25	−5	1.0815	0.5156	12.76	20.80	0.8496	0.4941	12.92	19.52
	5	1.0785	0.5508	12.56	16.18	0.8472	0.5057	12.98	18.36
	10	1.0815	0.9736	9.64	22.34	0.8643	0.3812	10.21	16.52

1.0804 and 1.0814 is obtained for image 1 and image 2 respectively. It also decreases with increase in angle like MI. Similarly when at same scaling if image is rotated anti-clockwise NMI at −15 is 1.077 for image 1 and 1.0810 for image 2. Further, it decreases as image is rotated anti-clockwise. PSNR is maximum at angle 15 that is 11.24 and 11.28 for the images. The maximum value of PCC comes out to be 0.9736 and minimum is 0.3812.

4.4 Summary of Normalized Mutual Information

Normalized Mutual Information decreases with increase in angle clockwise and when angle is increased anti-clockwise NMI decreases. With increase and de crease in NMI all other parameters are affected accordingly. When there is increase in the Scaling we rotated the images at comparatively smaller angles as with increased scaling and big angle rotations NMI almost declines giving poor quality image. Comparison of MI and NMI along with other metrics has been tabulated in Table 3.

4.5 Results of Image Registration on Dataset

The present method aligns each of image pairs from data set [5] with high accuracy. The two images cannot be aligned either when the images truly do not overlap or when there is insufficient information to determine an accurate, reliable transformation between images. The results performed on set of different image pairs are obtained and average value of all the parameters have been tabulated in Table 3.

Table 3 Results obtained for dataset of images

Scale	Angle	MI	NMI	PCC	PSNR	ET
0.5	−20	0.8132	1.0724	0.7175	9.79	17.01
	−15	0.8057	1.0715	0.8053	10.68	17.75
	15	1.8080	1.0721	0.7695	10.37	18.56
	25	0.8098	1.0731	0.7369	10.05	12.08
0.75	−20	0.7220	1.0684	0.6540	9.41	14.32
	−10	0.7961	1.0695	0.7496	10.16	14.87
	10	0.8283	1.0743	0.7499	10.78	14.52
	20	0.8018	1.0746	0.7462	9.22	15.72
1.25	−10	0.7726	1.0741	0.7920	9.96	19.09
	−5	0.8105	1.0731	0.6507	10.76	17.95
	5	0.7988	1.07387	0.6392	10.48	15.25
	10	0.7712	1.0722	0.5556	10.49	18.80

5 Conclusion and Future Scope

A method for evaluating the robustness of Intensity Based Image Registration using the Similarity Measure Mutual Information and Normalized Mutual Information and other metrics such as PCC, PSNR and elapsed time has been presented. Using above method NMI incorporates maximum information into the similarity measure of the two images and hence results show that the registration by NMI is much better than that by MI. With increase in rotations value of MI and NMI decreases and with increase in scaling, value of MI and NMI increases. Results also indicate that the registration will not be affected with the increase in robustness of NMI and MI based methods. In future, the present work may be extended to investigation and development of a better measure for robustness of Intensity based registration methods. Non-rigid image registration can also be presented by applying MI and NMI based registration algorithms.

References

1. A. Hacine Gharbi, M. Deriche, P. Ravier, R. Harba, T. Mohamadi, A new histogram based estimation technique of entropy and mutual information using mean squared error minimization. Comput. Electr. Eng. **39**(3), 918–933 (2013)
2. L.G. Brown, A survey of image registration techniques. ACM Comput. Surv. **24**(4), 325–376 (1992). https://doi.org/10.1145/146370.146374. http://doi.acm.org/10.1145/146370.146374
3. P.A. Legg, P.L. Rosin, D. Marshall, J.E. Morgan, Feature neighbourhood mutual information for multi-modal image registration: an application to eye fundus imaging. Pattern Recogn. **48**(6), 1937–1946 (2015)
4. M.A. Viergever, J.A. Maintz, S. Klein, K. Murphy, M. Staring, J.P. Pluim, A survey of medical image registration–under review. Med. Image Anal. **33**, 140–144 (2016)

5. B.D. de Vos, F.F. Berendsen, M.A. Viergever, H. Sokooti, M. Staring, I. Išgum, A deep learning framework for unsupervised affine and deformable image registration. Med. Image Anal. **52**, 128–143 (2019)
6. Z. Gao, B. Gu, J. Lin, Monomodal image registration using mutual information based methods. Image Vis. Comput. **26**(2), 164–173 (2008)
7. G. Balakrishnan, A. Zhao, M.R. Sabuncu, J. Guttag, A.V. Dalca, An unsupervised learning model for deformable medical image registration. in *Proceedings of the IEEE Conference on Computer Vision and Pattern Recognition*, (2018), pp. 9252–9260
8. G. Reeves, H.D. Pfister, A. Dytso, Mutual information as a function of matrix snr for linear gaussian channels. in *2018 IEEE International Symposium on Information Theory (ISIT)*, (IEEE, 2018), pp. 1754–1758
9. A. Kelman, M. Sofka, C.V. Stewart, Keypoint descriptors for matching across multiple image modalities and non-linear intensity variations. in *2007 IEEE Conference on Computer Vision and Pattern Recognition*, (IEEE, 2007), pp. 1–7
10. X. Liu, Z. Duan, W. Xu, Improved computing method of mutual information in medical image registration. Int. J. Sig. Process. Image Process. Pattern Recogn. **9**(4), 415–424 (2016)
11. S. Jain, N. Kanwal, Overview on image registration. in Medical Imaging, m-Health and Emerging Communication Systems (MedCom), 2014 International Conference on, pp. 376–381. IEEE (2014)
12. F. Maes, A. Collignon, D. Vandermeulen, G. Marchal, P. Suetens, Multimodality image registration by maximization of mutual information. IEEE Trans. Med. Imaging **16**(2), 187–198 (1997)
13. Poole, Ben, Sherjil Ozair, Aaron van den Oord, Alexander A. Alemi, and George Tucker. "On variational bounds of mutual information." arXiv:1905.06922(2019)
14. Saxena, S., Singh, R.K.: A survey of recent and classical image registration methods. International Journal of Signal Processing, Image Processing and Pattern Recognition (2014)
15. C.E. Shannon, A mathematical theory of communication. ACM SIGMOBILE Mobile Computing and Communications Review **5**(1), 3–55 (2001)
16. P. Viola, W.M. Wells III, Alignment by maximization of mutual information. Int. J. Comput. Vision **24**(2), 137–154 (1997)
17. McCarthy, Arya D., and David W. Matula. "Normalized mutual information exaggerates community detection performance." In *SIAM Workshop on Network Science, SIAM NS*, pp. 78–79. 2018
18. T.O. Kvålseth, On normalized mutual information: measure derivations and properties. Entropy **19**(11), 631 (2017)
19. K.Z. Szabó, G. Jordan, A. Petrik, Á. Horváth, C. Szabó, Spatial analysis of ambient gamma dose equivalent rate data by means of digital image processing techniques. J. Environ. Radioact. **166**, 309–320 (2017)
20. J. Woo, M. Stone, J.L. Prince, Multimodal registration via mutual information incorporating geometric and spatial context. IEEE Trans. Image Process. **24**(2), 757–769 (2015)
21. H. Luan, F. Qi, Z. Xue, L. Chen, D. Shen, Multimodality image registration by maximization of quantitative–qualitative measure of mutual information. Pattern Recogn. **41**(1), 285–298 (2008)
22. Q.R. Razlighi, N. Kehtarnavaz, S. Yousefi, Evaluating similarity measures for brain image registration. J. Vis. Commun. Image Represent. **24**(7), 977–987 (2013)

A New Contrast Based Degraded Document Image Binarization

Usha Rani, Amandeep Kaur, and Gurpreet Josan

Abstract The binarization of badly degraded document images is very challenging job due to the presence of various degradations such as ink bleed through, stains, smear, blur, low contrast and nonuniform illumination. Many binarization techniques are proposed in the literature, most of these techniques are threshold based. This chapter proposes the binarization technique which uses the contrast feature to compute the threshold value with minimum parameter tuning. It computes the local contrast image using maximum and minimum pixel values in the neighbourhood. The high contrast text pixels in the image are detected using global binarization. Finally, the local thresholds are computed using high contrast image pixels within the local window to binarize the document image. It has been tested on benchmark datasets H-DIBCO-2010 and H-DIBCO-2016 in terms of F-measure, PSNR and NRM. The results are compared with the Bernsen's and LMM contrast based binarization techniques and found to be outperforming these methods.

Keywords Binarization · Contrast image · Degraded documents · Pre-processing

1 Introduction

Binarization of document images is one of the most important pre-processing steps of the document image analysis systems. It converts gray scale images into binary images to reduce the computational overload of the image processing systems. The performance of next phases such as segmentation and recognition are entirely dependent upon the binarization result. Due to various types of degradations present in

U. Rani (✉)
University College, Ghanaur, Patiala, Punjab, India
e-mail: usha_gupta7@yahoo.co.in

A. Kaur
Central University of Punjab, Bathinda, Punjab, India

G. Josan
Department of Computer Science, Punjabi University, Patiala, India

© Springer Nature Switzerland AG 2020
P. K. Mallick et al. (eds.), *Cognitive Computing in Human Cognition*,
Learning and Analytics in Intelligent Systems 17,
https://doi.org/10.1007/978-3-030-48118-6_8

the document images such as ink bleed through, stains, smear, blur, low contrast and nonuniform illumination, it is very difficult to get proper binarization results of the degraded document images. Many binarization techniques are proposed in the literature, most of these techniques are threshold based. Broadly, the binarization techniques are categorized as global and local thresholding methods. Global methods use single threshold value to binarize the entire image. Global methods are very fast but these are not suitable to binarize the document images which do not have bimodal histogram. Otsu [1] is most successful binarization method among all global techniques. Local binarization methods estimate a local threshold for each pixel in the document image, so these can handlethe high variation of intensities due to degradations in the document images and produce better results compared to the global methods. Niblack [2], Welner [3], Sauvola and Pietaksinen [4], Wolf and Jolion [5], Khurshid et al. [6], are well known local adaptive methods of image binarization which utilize the statistical measures such as mean, standard deviation or variance in the neighborhood of the pixel under consideration to determine the threshold value. Bernsen [7] and Su et al. [8] compute local contrast using maximum and minimum intensity values in the neighborhood. Gatos et al. [9], Zhou et al. [10] and Kawano et al. [11] are local adaptive binarization methods which are based on the background estimation and subtraction. Kim et al. [12], Oh et al. [13] and Valizadeh and Kabir [14] are water flow model based local binarization techniques. Jiangtaoet al. [15] binarizeduneven illuminated images using curvelet transform and Otsu's method. Elazab et al. [16] proposed adaptively regularized kernel-based fuzzy C-means clustering method for segmentation of brain magnetic resonance images. Kernel fuzzy c-means (KFMC) method proposed by Farahmand et al. [17] performs noise removal and binarization simultaneously by classifying image pixels into foreground, background and noise. A region based adaptive binarization technique for unevenly illuminated images is proposed by Michalak and Okarma [18]. Ouafek and Kholladi [19] presented multistage binarization technique using Artificial Neural Network.

This paper proposes a local adaptive binarization method based on local contrast is the modified form of LMM [8], which computes the local threshold image using maximum and minimum pixel values in the neighbourhood. The high contrast text pixels around the text stroke boundary are detected using global binarization. Finally, local threshold values are computed using high contrast image pixels to binarize the document image.

The rest of the paper is organized as follows: The work related with the proposed technique is discussed in Sect. 2. The proposed binarization technique is discussed in Sect. 3. The experimental results are given in Sect. 4. Finally, Sect. 5 gives the conclusions.

2 Related Work

This section describes the work related with the proposed degraded document image binarization technique.

Bernsen's Method [7] is the simplest contrast-based thresholding method that computes local contrast $C(i,j)$ using difference between the highest and lowest pixel gray values in the local window around pixel (i,j) of the considered image as follows:

$$C(i, j) = I_{max}(i, j) - I_{min}(i, j) \tag{1}$$

The pixel is classified into text or background according to the local contrast is above or below the user-provided contrast threshold (CTh). If the local contrast $C(i, j)$ is below the contrast threshold (CTh), the pixel is set as background, otherwise the pixel is classified into text or background depending on the value of the local mid-grey value that is $(I_{max}(i, j) + I_{min}(i, j))/2$. This method is computationally simple as well as fast as compared to other local binarization techniques which involve mean, variance, histogram calculations etc. The problems with this method are that results are very much dependent upon the neighborhood size and it does not work properly on degraded images with a complex background.

Su et al. [8] method also known as local maximum minimum method (LMM) based on local contrastis an improvement over Bernsen's method and handles the document images with a complex background well. This method computes the local contrast using window size 3×3 as follows:

$$C(i, j) = \frac{I_{max}(i, j) - I_{min}(i, j)}{I_{max}(i, j) + I_{min}(i, j) + e} \tag{2}$$

$I_{max}(i, j)$ and $I_{min}(i, j)$ represent the maximum and minimum pixel values in the window centered at (i, j). A positive but infinitely small number e is added in the denominator in case the $I_{max}(i, j)$ is equal to zero. Normalization factor (the denominator) is introduced in Eq. (2) to compensate the effect of image contrast and brightness variation.

If the number of high contrast pixels (N_e) with in the local window is greater than a threshold N_{min}, then document image pixel is considered as the text pixel.

$$B(x, y) = \begin{cases} 1, & (N_e \geq N_{min}) AND(I(i, j) \leq E_m + E_\sigma/2) \\ 0, & \text{otherwise} \end{cases} \tag{3}$$

Here,

$$E_m = \frac{\sum_{neighbor} I(i, j) \times (1 - E(i, j))}{N_e} \tag{4}$$

$$E_\sigma = \sqrt{\frac{\sum_{neighbor}((I(i, j) - E_m) \times (1 - E(i, j))^2}{2}} \tag{5}$$

In these equations, 'I' is the input document image, E is the binary high contrast pixel image in which high contrast pixel values are zero. The image E is obtained by binarization of local contrast image (C) using Otsu method. N_e is the number of high contrast pixels within the local window. The local window size and minimum number of high contrast pixels within the local window (N_{min}) are the parameters that are needed to set. Authors related these parameters to the text stroke width in the document image under study.

3 Proposed Work

This section describes the proposed binarization technique which is the modified form of LMM method. LMM method need two parameters, one is window size and second is N_{min} which are estimated from the stroke width. The method described below needs only one parameter window size, which is set equal to the stroke width.

1. The high contrast image using Eq. (2) results high contrast response at either side of text stroke edges, we computed contrast image using window size 3×3 as follows:

$$C(i, j) = \frac{I_{max}(i, j) - I(i, j)}{I_{max}(i, j) + e} \tag{6}$$

This equation reduces the background variation and assigns more accurate contrast value to document pixels than the contrast image created by Eq. 2 [20] as shown in Fig. 1.

2. The binary high contrast pixel image (E) is obtained by binarization of local contrast image (C) using Otsu method.

3. After this, the foreground text is extracted within sliding window of size equal to the stroke width using following formula:

$$I_b = \begin{cases} 0, & I(i, j) \leq E_m + \frac{E_\sigma}{2} < 0 \\ 1, & otherwise \end{cases} \tag{7}$$

Here, E_m is the mean and E_σ is the standard deviation of intensity of stroke edge pixels in the window and are calculated using Eqs. (4) and Eq. (5) respectively.

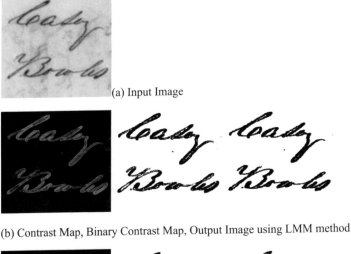

(a) Input Image

(b) Contrast Map, Binary Contrast Map, Output Image using LMM method

(c) Contrast Map, Binary Contrast Map, Output Image using proposed method

Fig. 1 Results of LMM method and proposed method

4 Experimental Results

The experiments are conducted to show the effectiveness of the proposed method using HDIBCO-2010 and HDIBCO-2016 dataset. These datasets contain a number of handwritten document images with different types of degradations. The results are compared with existing contrast-based techniques: Bernsen's method and LMM method. Figures 1 and 2 shows the visual results of sample of images taken from the datasets. The F-measure, PSNR, NRM are the quantitative metrics adapted from DIBCO's reportsto evaluate theperformance of the considered existing methods and the proposed method. The average values of F-measure, PSNR and NRM of all images in HDIBCO-2010 and HDIBCO-2016 dataset are recorded in Table 1 and Table 2 respectively, which shows that proposed method outperform the other considered methods. The value of F-measure and PSNR should be high and value of NRM should be low for good quality of binarization.

(a) Input Images

(b) Groundtruth Images

(c) Bersen's method results

(d) LMM method results

(e) Proposed method results

Fig. 2 Binarization results of input images with different methods

Table 1 Evaluations results of dataset HDIBCO-2010

Method	F-measure (%)	PSNR	NRM($\times 10^{-2}$)
Bernsen	41.5	8.57	21.18
LMM	85.49	17.83	11.46
Proposed method	**89.7**	**19.38**	**8.27**

Table 2 Evaluations results of dataset HDIBCO-2016

Method	F-measure (%)	PSNR	NRM($\times 10^{-2}$)
Bernsen	44,1	7.67	14.98
LMM	81.43	15.87	11.8
Proposed method	**83.5**	**16.63**	**11.2**

5 Conclusions

This paper proposes the binarization technique especially for handwritten degraded document images. This method first estimates the contrast image and then local thresholding is applied using the mean and standard deviation of detected high contrast pixels in the contrast image. Though the proposed method outperforms the other methods considered in this study, but it is unable to extract very low contrast pixels in the image so we can work to further improve this method.

References

1. N. Otsu, A threshold selection method from gray level histograms. IEEE Trans. Sys. Man Cybern. **9**(1), 62–66 (1979)
2. W. Niblack, *An Introduction to Image Processing* (Prentice-Hall, Englewood Cliffs, NJ, 1986), pp. 115–116
3. P. Wellner, Adaptive Thresholding for the Digital Desk. Xerox, EPC1993–110 (1993)
4. J. Sauvola, M. Pietaksinen, Adaptive document image binarization. Pattern Recogn. Lett. **33**(1), 225–236 (1997)
5. C. Wolf, J.M. Jolion, Extraction and recognition of artificial text in multimedia document. Pattern Anal. Appl. **6**(4), 309–326 (2003)
6. K. Khurshid, I. Siddiqi, C. Faure, N. Vincent, Comparison of Niblack inspired methods for ancient document. in *16th IEEE International Conference on Document Recognition and Retrieval 7247* (2009), pp. 1–10
7. J. Bernsen, Dynamic thresholding of gray level images. in *Proceedings of IEEE International Conference on Pattern Recognition* (1986), pp. 1251–1255
8. B. Su, S. Lu, C.L. Tan, Binarization of historical handwritten document images using local maximum and minimum filter. Int. Workshop Doc. Anal. Sys. pp. 159–165 (2010)
9. B. Gatos, I. Pratikakis, S.J. Perantonis, Adaptive degraded document image binarization. Pattern Recogn. **39**(3), 317–327 (2006)
10. S. Zhou, C. Liu, Z. Cui, S. Gong, An improved adaptive document image binarization method. in *Proceedings of 2nd IEEE International Congress on Image and Signal Processing* (2009), pp. 1–5

11. H. Kawano, H. Oohama, H. Maeda, Y. Okada, N. Ikoma, Degraded document image binarization combining local statistics. in *IEEE International Joint Conference (ICROS-SICE)* (2009), pp. 439–443
12. I.K. Kim, D.W. Jung, R.H. Park, Document image binarization based on topographic analysis using a water flow model. Pattern Recogn. **3**, 265–277 (2002)
13. H.H. Oh, K.T. Lim, S.I. Hien, An improved binarization algorithm based on a water flow model for document image with inhomogeneous backgrounds. Pattern Recogn. **38**, 2612–2625 (2005)
14. M. Valizadeh, E. Kabir, An adaptive water flow model for binarization of degraded document images. IJDAR **16**(2), 165–176 (2013)
15. W. Jiangtao, L. Shumo, S. Jiedi, A new binarization method for non- uniform illuminated document images. Pattern Recogn. **46**(6), 1670–1690 (2013)
16. A. Elazab, F. Jia, J. Wu, Q. Hu, Segmentation of brain tissues from magnetic resonance images using adaptively regularized kernel-based Fuzzy C-Means Clustering. in *Computational and Mathematical Methods in Medicine* (Hindawi Publishing Corporation, 2015)
17. A. Farahmand, A. Sarrafzadeh, J. Shanbehzadeh, Noise removal and binarization of scanned document images using clustering of features. in *Proceedings of the International MultiConference of Engineers and Computer Scientists*, vol. 1 (2017)
18. H. Michalak, K. Okarma, Region based adaptive binarization for optical character recognition purposes. In: International Interdisciplinary Ph.D. Workshop (IIPhDW) (Swinoujście, 2018) pp. 361–366
19. N. Ouafek, M. Kholladi, A binarization method for degraded document image using artificial neural network and interpolation inpainting. in *Proceedings of 4th International Conference on Optimization and Applications (ICOA)* (Mohammedia, 2018) pp. 1–5
20. B. Su, S. Lu, C.L. Tan, Combination of document image binarization techniques. in *International Conference on Document Analysis and Recognition* (Beijing, China, 2011) pp. 22–26

Graph Based Approach for Image Data Retrieval in Medical Application

Subrata Jana, Sutapa Ray, Pritom Adhikary, and Tribeni Prasad Banerjee

Abstract Now-days most of the people are suffering in diabetic, retinopathy and glaucoma. It is non-invensive action and experimental study. In general manual images are manually seen by trained clinician in a time consume and not accurate result. This paper describes new technique such as max-flow graph base approach. Max-flow based techniques more accurate result than PCA and Hessian method. Initial get fundus image on DRIVE and STAIR database, show that the new graph base technique is effective and positive in relation to similar System.

Keywords Convolution · Energy minimization · Histogram · Max-flow · Min-cut · Segmentation

1 Introduction

The computerized segmentation are detect eye vessel from fundus image. Blood vessel is minor extended structures in retina. Retinal vessel altered by abundant methodical diseases, such as diabetes, hypertensions and glaucoma. Organize an unorganized combined segmentation to detect different eye vessel. It is useful tool to accurate, fast judgment for final disease [1]. Retina and Brain segmentation are

S. Jana (✉)
MCA Department, Calcutta Institute of Technology, Howrah, West Bengal, India
e-mail: 2007.subrata@gmail.com

S. Ray
ECE Department, Institute of Engineering and Management, Kolkata, India
e-mail: rayadhikary.sutapa@gmail.com

P. Adhikary
EE Department, Calcutta Institute of Technology, Howrah, West Bengal, India
e-mail: pritom06@gmail.com

T. P. Banerjee
ECE Department, Dr. B. C. Roy Engineering College, Durgapur, West Bengal, India
e-mail: tribeniju@gmail.com

© Springer Nature Switzerland AG 2020
P. K. Mallick et al. (eds.), *Cognitive Computing in Human Cognition*,
Learning and Analytics in Intelligent Systems 17,
https://doi.org/10.1007/978-3-030-48118-6_9

91

related to pediatric cerebral malaria. We can easily identify similarities and dissimilarities retinal and cerebral data in retinopathy-positive pediatric cerebral malaria [2]. In diabetic patient retina vessel are bending but normal patient vessel does not bend. After segmentation we see vascular architecture and predict what type of patient. The retina is visible and available to high motion non-invasive image. Single window that allow express visualization and study of the inner retinal vessel for review different related condition [3]. There are two segment the eye vessel either manual vessel and automated segment the vessel. In manual vessel does not she slight vessel but automated segmentation technique visualized slight vessel and effected vessel [4]. vessels are filtering from 3D image it is more accurate and well compute best possible medical axes. It is more accurate result than other technique [5]. The system is found on removal of image ridge, which overlap approximately with vessel Center lines. The edge is used to make up primitives in the form of line component [6]. At the moment middle aged people vascular disease technique such as Coronary Heart Disease (CHD) [7]. Micro aneurysms detection one of the most important step automatic segmentation of Diabetic Retinopathy (DR). In fundus image capillary vessel are not visible but graph base segmentation capillary vessel are visible. Micro aneurysms are between the first sign of the event of diabetic patient. Blood vessels are drawing out from retinal background, identification the arteries and veins are finding and analysis of peculiar regions such as hemorrhage, exudates, optic disc, artery venous crossing [8]. There are so lots of method remove the vessel, so many techniques are morphological operation, clustering, snake, thresholding etc. But new tech is Graph base approach, such as B-spline, graph-cut, max-flow, min-cut etc. Some investigator had worked graph base technique [9]. Graph Transformation Matching (GTM) technique are ruling the conformity graph up-and-coming from supposes matches. GTM techniques results are overlapping regions, and no matching points between images In order to obtain better outlook and more healthy outcome sing more matching point to get healthier mapping approximation [10]. We talk in relation to the results of the first global micro aneurysm recognition contest, ready in the situation of the Retinopathy Online Challenge (ROC), a multilayer online contest for various aspects of DR recognition. FROC system graphically represents false positive and true positive recognition per image [3, 11]. Extract Vasculature from 2D medical images. Morphological operations extract the vessel using open and close operation. All data are store in cloud then analysis in health care area [12]. In recent times used Deep Learning is a new technology classify difficult type of disease such as diabetic, retinopathy, glaucoma etc. it also used different type sensor, IOT, smart phone application [13–19]. ELM technique are used for basic machine learning task. Machine learning technique easily identify different type disease [20–22]. Dense Sub graphs express region and narrow information about local graph therefore easily identify small vessel [10, 23]. Graph base application are linear application recently all linear base application to change nonlinear base application. Nonlinear base application are reducing size of data, dynamic network, and classification and clustering over dynamic [23]. Our wave propagation and trace back algorithm can be easily extended

Fig. 1 Min-cut/max-flow
theorem

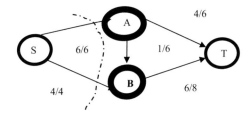

to operate 3D data. Section 2 Background of Graph Theory, Sect. 3 problem speci-
fication using graph. Section 4 Max-flow algorithm discusses. Section 5 Explain the
result. Section 6: Conclusion.

2 Background Graph Theory

Graph is combined put of vertices and edges. A graph is a couple $G = (V, E)$ of
put fulfilling $E \subseteq |V|^2$ thus, the element of E are element subsets of V. To stay
away from notational haziness, we shall always think tacitly that $V \cap E = \varphi$. The
essentials of V are the vertices (or node, or points) of the graph G, the elements of
E are its edges (lines). A biased graph directed graph has two workstation source
(S) and sink (T). We think about as flow network with edge capability resolve edge
cost that is weight. Edge can flow not greater than edge capacity. Max-flow min-cut
theorem flow is maximum equal to minimum cut that can be pass source S to Sink
T. In Fig. 1 maximum flow can be send from source S to Sink T range 10 is equal to
minimal cut $C = \{(S, A), (S, B)\}$.

3 Problem Specification Using Graph

Vessel and retina are two types' combines in eye. In Human body Heart are circulate
blood by blood vessel. Blood are circulated from heart by blood vessel. The noticeable
vascular formation is truly cycle free. The vessel can be notable by figure color and
texture quality. We can suggest the vessels as curvy segments S, which could division
and fractious. Planner graph $G(V, E)$ *are combined*, where an one to one each edge
E_i $(j < k \le m)$ keep up a correspondence to vessel segment S_j. The node $V_j (j < q \le$
$n)$ of the graph stand for the branches or equivalent to vessel segments. If two vessels
exasperated each and every one other both are separating into two vessel segments
keeps up a correspondence to by two edges. The nodes are equivalent to pixel. We
think about additional two nodes source and sink. Source (S) and Sink (T) has two
terminals in a graph.

Now a point denote foreground and b points background the statement can be
defined [24]

$$\max\{g\} = \sum_{i \in a} f_i + \sum_{i \in b} b_i - \sum_{i \in a, j \in b \vee j \in a, i \in b} z_{i,j} \tag{1}$$

Above maximization problem can be converted into minimization problem, the problem can be formulated is [24]

$$\min\{g\} = \sum_{i \in a, j \in b \vee j \in a, i \in b} z_{i,j} \tag{2}$$

in Eq. (2) denote minimum cut problem. Min-cut problem are constructing a network where the source is connected to the entire pixel with ability f_i, the sink is related by all the pixel with ability n_j. Two edges (l, j) and (j, i) with z_{ij} capability are added between two contiguous pixel. There are two separate sets of vertices V_f and V_b that represent the source and sink of the graph, such that the source S is connected to each node V_f, while each node of V_b is connect to the sink t. We can define them [24]

$$V_f = \{(m', n', 0) : 0 \le m' \prec m'size, 0 \le n' \prec n'size\}$$
$$V_b = \{(m', n', o'size) : 0 \le m' < m'size, 0 \le n' > b'size\} \tag{3}$$

The minimum cut C_{min} obtained by computing the maximum flow over G Contains a set of edges of minimum total capacity that isolates the source and sink. C_{min} is defined as [24]

$$\arg\min_d \left(\sum_{(a1,b1) \in d} d(a1, b1) \right)$$
$$= \arg\min_d \left[\sum_{(a1,b1) \in d_{label}} d(a1, b1) + \sum_{(a1,b1) \in d_{penalty}} d(a1, b1) \right] \tag{4}$$

where $D_l = D \cap F_l$ and $D_p = D \cap F_p$.

4 Max-Flow Algorithm Explanation

Max-flow algorithm construct directed graph and create directed path from source to sink. Source node denoted foreground and sink node denoted background. The max-flow algorithm frequently keep up three stages:

1. Growth stage: Source (p) and sink (q) are increasing until create a path p → q.
2. Expansion stage: tree create the path source (p) to sink (q), tree partition into small subtree (p).
3. Adaptation stage: tree return p and q.

When construct directed path form source to sink in between three node are created (1) Active, (2) Passive and (3) free node. Active node allow to grow search tree for new children and passive node can't grow search tree such as completely block the node. Active node search next active node, if we found next active node search tree are expanded otherwise block the search tree. Active node maintains two queues. Active node are new to the end of the second queue and read from the front of the first queue, if first queue is empty, it is replace by second queue. There are two type of queue queue_last and queue_first. Now we explain set_Active node algorithm:

set_Active(node * head):
step 1: head_next <> NULL
step 2: queue_last → next = head
step 3: queue_first = head
step 4: return next node

Now we explain next_Active_node() algorithm:

step 1: node parent
step 2: while(1)
Remove from active list
if(parent-next = parent)
queue_first = queue_last = NULL
step 3: queue_first = parent-next
step 4: return parent

at the present we clarify augmentation stage algorithm:
The input for this phase is a pathway m from source p to sink q. At the aperture orphan_set is vacant but a number of orphans in the finish since at smallest quantity one edge in m be transformed into flooded.

Ruling the traffic jam capability ω on m
Modify the wonderful diagram with practically flow ω through m
for both edge(m, n) in m that converted into drenched.
  queue_first(m) = queue_last(n) = s then set neighbournode(n) = ψ and or = or
$\cup$ {n}
  queue_first(m) = queue_last(n) = t then set neighbournode(n) = ψ and or = or
$\cup$ {m}

Now we explain adaptation stage:

every node m is frustrating to locate new-fangled valid parent in search tree and first search the new-fangled parent node otherwise it develop into a free node and all the children are joined.

while or <> ψ
choose an orphan node m $\in$ or remove from or
process m
end while

The process m consist keep up some steps. First we are irritating to locate suitable parent m amongst its neighbors. Neighboring node search suitable root node and added adaptation list. A suitable root n should assure: queue_last(n) = queue_first(m), tr_cap(n → m) > 0 and the starting point of the n starting place node. We can identify starting place node set marks along path with neighboring node.

The procedure process m consist maintain some steps. First we are annoying to find valid root m between its neighbors. Neighboring node search suitable root node and added adaptation list. A suitable root node m should satisfy:queue_first(m) = queue_last(n), tr_cap(m → n) < 0 and the terminal of the m sink node. We can identify sink node set marks along path with reverse neighboring node.

5   Experiment Result

In Fig. 2 there are different images get from DRIVE database. We can use three different, techniques such as (1) PCA, (2) Hessian, (3) max-flow. In PCA base technique thing vessel are not showing. In Hessian base technique does not clear result. In max-flow base technique clear result and all thing vessels are showing.

6   Conclusion

In this paper, we discuss new method for narrow vessel detection. The proposed method takes the relevant results of the earlier graph base application. Max-flow graph base technique much better result PCA and Hessian Technique. All Image collected from DRIVE and STAIR database.

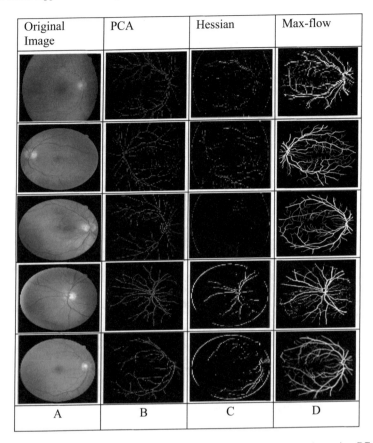

Original Image	PCA	Hessian	Max-flow
A	B	C	D

Fig. 2 Vessel segmentation technique **a** original image, **b** vessel segmentation using PCA base technique, **c** vessel segmentation Hessian base technique, **d** vessel segmentation using max-flow base technique

References

1. M.M. Fraz, P. Remagnino, A. Hoppe, B. Yyanoonvara, A.R. Rudricka, C.G. Owen, S.A. Barman, Blood vessel segmentation methodologies in retinal images—a survey. Comput. Methods Programs Biomed. **108**, 407–433 (2012)
2. Y. Zheng, M.T. Kwong, I.J.C. MacCormick, N.A. Beare, S.P. Harding, Comprehensive texture segmentation framework for segmentation of capillary non-perfusion region in fundus fluoresce, in angiogram. PLOS ONE 0093624 (2014)
3. M. Niemeijer, B.V. Ginneken, M.J. Cree, A. Mizutani, G. Quellec, B. Zhang, R. Hornero, W.U. Xiangqian, G. Cazuguel, A. Mayo, M.D. Abramoff, Retinopathy online challenge: automatic detection of microaneurysms in digital color fundus photographs. IEEE Trans. Med. Imaging **29**(1), 185–195 (2010)
4. M. Niemeijer, J. Stall, B.V. Ginneken, M. Loog, M.D. Abramoff, Comparative study of retina vessel segmentation methods on a new publicly available database. SPIE Med. Imaging **5370**, 648–656 (2004)

5. M. Poon, G. Hamameh, R. Abugharbieh, Live vessel extending live wire for simultaneously extraction of optimal medical and boundary paths in vascular images, in *Medical Image Computing and Computer Assisted Intervention*, pp. 444–451 (2007)

6. J. Staal, M.D. Abramoff, M. Niemeijer, M.A. Viergever, B.V. Ginneken, Ridge based vessel segmentation in color images of retina. IEEE Trans. Med. Imaging 23(4), 501–509 (2004)

7. J. Wang, G. Liew, T. Wong, W. Smith, R. Klein, S. Leeder, P. Mitchell, Retinal vascular caliber and the risk of coronary heart disease related death. Heart 92(11), 1583–1587 (2006)

8. K. Akita, H. Kuga, A computer method of understanding ocular fundus image. Pattern Recogn. 15(6), 431–443 (1982)

9. W. Aguilar, M.E. Martinezperez, Y. Frauel, F. Escolano, M.A. Lozano, A.E. Romero, Graph base methods for retinal mosaicing and vascular characterization, in *Pattern Recognition 6th IAPR_TC-15, International Workshop*, pp. 25–36 (2007)

10. Y. Boykov, V. Kolmogorov, An experimental comparison of min-cut/maxflow algorithm for energy minimization vision. IEEE Trans. PAMI 26(1), 1124–1137 (2004)

11. F.K.H. Quek, C. Kirbas, Vessel extraction in medical images by wave propagation and traceback. IEEE Trans. Med. Imaging 20(2), 117–131 (2001)

12. J. Luo, M. Wu, Y. Zhao, Big data application in bio medical research and health care: a literature review. Biomed. Inform. Insights 8, 1–10 (2016). https://doi.org/10.4137/bii.s31559

13. W. Farhan, Z. Wang, Y. Huang, S. Wang, F. Wang, X. Jiang, A predictive model for medical events based on contextual embedding of temporal sequence. JMIR Med. Inform. 4(4), e39 (2016)

14. N. Razavian, J. Marcus, D. Sontang, Multi-task prediction of disease onsets from longitudinal lab tests. JMLR W&C 56, 73–100 (2016)

15. E. Nurse, B.S. Mashford, A.J. Yepes, Decoding EEG and LFP signals using deep learning: heading TrueNorth, in *ACM International Conference on Computing Frontiers*, pp. 259–266 (2016)

16. D. Ravi, C. Wong, B. Lo, G.Z. Yang, A deep learning approach to on-node sensor data analytics for mobiles or wearable devices. IEEE J. Biomed. Health Inform. 21(1), 56–64 (2017)

17. P. Nguyen, T. Tran, N. Wickramasinghe, S. Venkatesh, Deepr: convolutional net for medical record. IEEE J. Biomed. Health Inform. 21, 22–30 (2017)

18. R. Miotto, F. Wang, S. Wang, X. Jiang, J.T. Dudley, Deep learning for healthcare: review, opportunities and challenges. Brief. Bioinform. 1–11 (2017). https://doi.org/10.1093/bib/bbx044

19. M. Voets, K. Mollersen, L.A. Bongo, Replication study: development and validation of a deep learning algorithm for detection of diabetic retinopathy in retinal fundus photograph. arXiv: 1803.04337v3(cs.cv) (2018)

20. F. Nie, H. Wang, C. Deng, X. Gao, X. Li, H. Huang, New l1-norm relaxations and optimizations for graph clustering, in *30th AAAI Conference on Artificial Intelligence*, pp. 1962–1968 (2016)

21. K. Hu, W. Wang, X. Gao, Microcalcification diagnosis in digital mammography using extreme learning machine based on hidden Markov tree model of dual tree complex wavelet transform. Expert Syst. Appl. 86, 135–144 (2017). https://doi.org/10.1016/j.eswa.2017.05.062

22. T. Liu, C.K.L. Lekamalage, G.B. Huang, Z. Lin, An adaptive graph learning method based on dual data representations for clustering. Pattern Recogn. 77, 126–139 (2018)

23. A.L. Agasta Adline, G.S. Mahalakshmi, S. Sendhilkumar, Graph based generation of research paper summaries. J. Comput. Theoret. Nanosci. 15(4), 1106–1111 (2018). https://doi.org/10.1166/jctn.2018.6567

24. S. Roy, Stereo without epipolar lines: a maximum-flow formution. IJCV 34(2/3), 147–161 (1999)

Brain Computer Interface: A New Pathway to Human Brain

Poonam Chaudhary and Rashmi Agrawal

Abstract The evolution of brain computer interface started with the need of subject's disability of verbal or written communication or to control immediate environment. Now days this field has been expanded other than neuroprosthetics applications and includes eminent areas of research like education, communication, entertainment, marketing and monitoring. This chapter focus on past 15 years, this assistive technology has attracted potentials numbers of users as well as researchers from multidiscipline.

Keywords BCI · SNR · GABA · SSVEPs

1 Introduction

The growth in the BCI research groups, journals, conferences, articles and number of attendees are evidences of the speedy growth the research field. Apart from these evidences, numerous projects are approved by different companies to develop BCI related applications. They also have announced their roadmaps to collaborate with different research groups for the development of BCI-based applications.

There are many annual conferences, workshops and seminar, which transmit latest developments in the field and give platform to prominent scientists to present their research projects such as National Center for Medical Rehabilitation Research of the National Institute of Child Health and Human Development of the National Institutes (USA), international conferences on Multimodal Interaction (ICMI), the IEEE/ACM International Conference on Computer-Aided Design (ICCAD), Intelligent User Interfaces (IUI), IEEE Transactions on Neural Systems and Rehabilitation Engineering, Journal of Neural Engineering etc.

P. Chaudhary (✉)
The NorthCap University, Gurugram, India
e-mail: poonam.potalia@gmail.com

R. Agrawal
Manav Rachna International Institute of Research and Studies, Faridabad, India
e-mail: drrashmiagrawal78@gmail.com

© Springer Nature Switzerland AG 2020
P. K. Mallick et al. (eds.), *Cognitive Computing in Human Cognition*,
Learning and Analytics in Intelligent Systems 17,
https://doi.org/10.1007/978-3-030-48118-6_10

99

An incursion of researchers from assorted disciplines, including rehabilitation, psychology, computer science, mathematics, medical physics, neurology and neuro-surgery and biomedical engineering is the justification behind the unusual growth of BCI research. Brain-Computer Interface is at the Innovation Trigger stage of the emerging technology mega-trends in the Gartner's 2018, 2017 and 2016 Hype Cycle. The predictions in the Gartner's Hype Cycle suggest that mainstream embracing will occur in more than 10 years for BCI research. This phenomenon is captured in Fig. 1.

The requisite knowledge of complex BCI designing involves BCI modes of oper-ation, experimental strategy, signal recording, types of measurable brain signals and feedback system [1–6]. The type of BCI can be divided on the basis of their mode of operation like synchronous or asynchronous, exogenous or endogenous. An exoge-nous BCI uses brain signals generated by the brain in the presence of external stimuli like visual or auditory stimuli that can elicit large response in the form of neuron activity. Steady State Visual Evoked Potentials (SSVEPs) and P300 are the example of control signals used by the exogenous BCI. Therefore the response of the exoge-nous is spontaneously generated brain patterns which don't require extensive user training. The advantages of such systems are less minimal training to user, single channel recording, easy and quick set-up of control signals, high information transfer rate. However user has to be more focused during the training phase which may

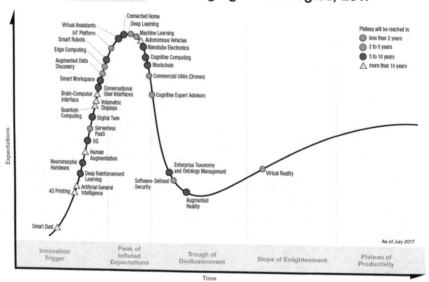

Fig. 1 Gartner's Hype Cycle

cause tiredness, fatigue. In contrast with exogenous BCI, an endogenous BCI uses the self-regulated brain rhythms and potentials generated in the brain without any external stimuli. User needs extensive neurofeedback training to learn to generate specific brain patterns. This category of BCI is directly dependent on the user's will and capability of learning the patterns. The endogenous BCI is beneficial for cursor control application using brain activity and for the users with sensory disabilities. The other criteria for the bifurcation of BCI systems is input data processing modality i.e. synchronous or asynchronous BCI. Synchronous BCI systems are the cue-based systems in which first set of features extracted and processed, then only another set of features are allowed to be extracted and processed. A predefined time window is decided and the signals belongs to that window are analyzed first. This system allows user to send commands only in predefined time frame. Regardless of user ability of modulating his/her brain signals, early and accurate detection of the control task can be acquired by using cueing process. This results into increase in confidence, sovereignty and interest of the user while taking the training of BCI skills. Beside easy and simple designing and evaluation of synchronous BCI as compare to asynchronous BCI, synchronous BCI is not very helpful in real world set-ups. However asynchronous or non-cue based BCI offers more practical approach for human-computer interaction. It does not require any sequence to extract and process the feature set. There is no predefined time frame for accepting and processing the feature set. User can act more normal and can initiate the communication by his/her will. It is also known as self-pace BCI. Independent of cue, this BCI system continuously analyzes the user's brain activity which leads it to real world set-ups.

Invasive BCI uses surgical implantation of microelectrode arrays inside the grey material of brain. Electrocortigography (ECoG) and Intracortical Neuron recording are the two invasive modalities in BCI research. Furthermore, in electrocortigography or intracortical neuron recording microelectrodes are placed on the surface of cortex. It could be Epidural Electrocortigography in which electrodes placed outside the dura mater or Subdural electrocortigography in which electrodes placed under the dura mater [1–6]. On the other hand, Intracortical neuron recording places the microelectrodes inside the cortex. Both the modality involves significant risk of infection and tissue damage in brain. Also scar-tissue build-up leads to issues related to long term stability. Though invasive modality leads to reasonable risk, it provides high quality of signals, very good spatial resolution and a higher frequency range.

Non-invasive BCI does not require any excruciating surgical procedure. The electrical activity generated by the millions of neuron can be recorded by placing small disc shape sensors known as electrodes on the scalp. This conventional and cost effective method has been used successfully in clinical and BCI research settings. It records signals at good temporal resolution i.e. change in signals within a specific time interval. However, the spatial resolution and frequency range is limited due to brain and non-brain artifacts. This results in the decrease in signal to noise ratio (SNR) as the frequency increases.

The brain computer interface is not a solitary mission. This vast multidiscipline endeavor includes neurology, concepts of instrumentation engineering and brain activity measurements, signal processing, computer science algorithms and statistical methods for brain activity pattern identification, training and feedback to the user.

2 Brain Anatomy

Most imperative part of the BCI systems is human brain. With advancement in the neuroscience researches, researchers are able to describe the complex structure and functions of the human brain. It is indispensable to know the anatomy of human brain, its different activities, measurable signals and prerequisite of BCI design.

2.1 Essential Brain Anatomy Brief

The human's central nervous system is consists of brain and spinal cord. The peripheral nervous system connects the central nervous system to rest of the body. Human brain is the center of the whole body. It gives the instruction to other body parts like sensory organs, other organs, muscles, glands, blood vessels through peripheral nervous system. The anatomy of brain divides the brain into cerebrum, cerebellum, and brainstem. The largest part of the brain is cerebrum which is composed of left and right hemispheres. Both of the hemispheres are connect to each other via corpus callosum (collection of white matter fibers). These hemispheres are further divided into four lobes known as: the frontal lobe, the parietal lobe, the occipital lobe and temporal lobe. The different responsibilities of these lobes are given in Table 1. Interpreting touch, vision and hearing, speech, reasoning, emotions, learning, and fine control of movement is associated with different locations on the cerebrum. Maintaining the body balance and body posture, coordination of muscles movements are the functions of cerebellum. It is located under the cerebrum. The last but not the least is brainstem which connects the cerebrum and cerebellum to the spinal cord. The heart rate, breathing body temperature, digestion, sneezing, wake-up and sleep cycles, vomiting, swallowing, coughing are main functions of brainstem.

Billions of neurons in human brain connected via thousands of synapses generate an electrochemical pulse called as action potential. This potential can be measured as electrical waveform known as brain wave or brain rhythm. These brain waves transmit the information via a specialized connection synapse to neighboring neuron which is received through dendrites connected to that neuron. In this way brain forms a dynamic neural network every time brain experience new facts or new remembered event. This network grows stronger with increase of transmission of signals between the neurons. Other than electrical signals, human brain contains thousands of neurotransmitters molecules in vesicles of axon, which amplify relay and modulate signals between neurons. Glutamate, GABA, acetylcholine, dopamine, adrenaline, histamine, serotonin and melatonin are some common neurotransmitters of human

Table 1 Different responsibilities human brain lobes

Brain component	Functions
Cerebrum: frontal lobe	• Personality, behavior, emotions • Reasoning, judgment, planning, problem solving • Speech: speaking and writing • Movement, planning • Intelligence, concentration, self awareness
Parietal lobe	• Interpretation of language, words • Sense of touch, pain, orientation • Interprets signals from vision, hearing, motor, sensory and memory, recognition • Perception
Occipital lobe	• Vision interpretation (processing of colors, light, movement) • Integrates visual experiences
Temporal lobe	• Understanding and interpretation of auditory stimuli • Language understanding, parts of speech • Memory • Organization and sequencing
Cerebellum	• Also known as "little brain" • Maintaining the body balance and body posture • Coordination of muscles movements
Brainstems	• Connects cerebrum and cerebellum to spinal cord • Heart rate, breathing body temperature, digestion, sneezing, wake-up and sleep cycles, vomiting, swallowing, coughing

brain. These chemical messengers help the brain wave to travel through neurons and information transmission is between the neurons achieved with the help of chemicals.

3 Brain Computer Interface

In 1999, First International meeting on Brain Computer Interface technology [5] took place in USA with 50 participants from 22 research group. BCI taxonomy, methods and approaches had proposed in review. Two main following approaches had discussed: (1) Operant Conditioning Approach, (2) Pattern Recognition Approach. Former approach considers the self-regulation of brain potentials or rhythms. The thought-translation device (TTD) developed in 2003 by authors [3] was based on self-regulations slow cortical potentials (SCP). The author's Wolpaw et al. [7] also used the self-regulations of brain rhymes for BCI. In this approach, no stimuli is present to user and user should know the real time feedback, enforced correct behavior according to the feedback and right training to user [8]. The later approach i.e. pattern recognition approach for BCI uses different mental task which activate potentials at specific cortical area of brain. These mental tasks include motor imagery tasks, arithmetic baseline tasks, visual tasks, and speech and emotion task. Different mental

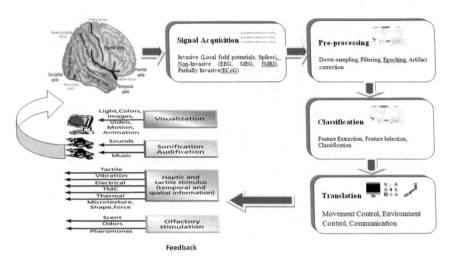

Fig. 2 Typical framework of brain computer interface

task activate the different patterns in EEG closed to the cortical areas detectable by scalp electrodes. Many BCIs [9–13] are based on this approach.

3.1 BCI Components

Figure 2 demonstrates the typical framework of brain computer interface comprising signal acquisition, pre-processing of acquired signals, feature extraction and selection, classification of these features into control actions and finally feedback to user for training of their minds. The orchestration of these components decides the performance measure of whole brain computer interface. The feature extraction, feature selection and classification can be replaced by deep learning algorithms too [14]. The following section will demonstrate each step in detail.

3.1.1 Signal Acquisition

There are different types of signals comprise of thermal, mechanical, electrical, chemical metabolic and magnetic activities inside the human brain generated due to intrinsic ignition. These signals can be recorded and become basis for alternative modes of communication and control. As discussed earlier, brain signals can be acquired by three methods (1) Non-invasive, (2) Partially invasive and (3) Invasive acquisition of signals. Figure 3 demonstrates positioning of electrodes on human brain according to acquisition method. Only non-invasive method does not involve any surgical procedure while others requires surgical procedure to place the electrode

Fig. 3 Brain's electrical activity acquisition methods

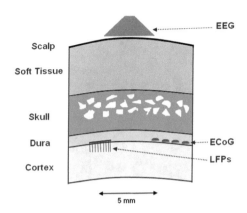

inside the skull. Scalp Electroencephalography (EEG), epidural electrodes and electrocorticography (ECoG), Local Field Potentials (LFPs), intracranial EEG (iEEG) are different methods to measure the electrical activity of human brain. Magnetoencephalography (MEG) [15] is the neuroimaging technique to measure magnetic fields produced by the electrical activity of the brain. The blood flow inside the brain also creates the neural activity which again can be imaged using functional magnetic resonance imaging (fMRI) and positron emission tomography (PET). Magnetic resonance spectroscopy (MRS) measures the chemicals (neurotransmitters) produces by the neural activity of brain. Invasive and non-invasive are two approaches of acquiring the brain signals [1–6].

Electroencephalograph (EEG)

Among all the various methods, EEG is most explored and experimented method for BCI systems. Electroencephalography is a non-surgical method used for measuring the electrical activity generated inside the brain. The temporal resolution of EEG is in milliseconds or better which is very good in terms of signal processing. But the spatial resolution is poor and in the range of centimeters. Spatial resolution depends upon the number of electrodes placed on the scalp. The position of electrodes also referred as channel and the distance between these channels is in few centimeters. The available EEG recording cap uses maximum 256 channels for recording. The amplitude and frequency are two basic features to characterize the EEG signals. The amplitude of EEG signals vary between 10 and 100 μV and frequency ranges between 10 and 1000 Hz. EEG patterns can be tracked above 256 Hz sampling rate and its frequency component ranges approximately between 10 and 100 Hz [16–19]. Figure 4 gives a glimpse of International 10/20 Standard for 64 + 2 channels EEG placement positions [20] for signal acquisition.

The electrical activity never stops as brain remains active always even when one is in sleep or unconsciousness. However, it does not mean that there would be general patterns. Brain waves are so irregular most of the time. According to Allison [21]

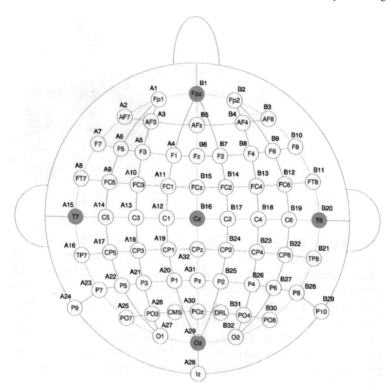

Fig. 4 International 10/20 standard for 64 + 2 channels EEG placement positions [20]

activity of any neural network makes a pattern or visible in EEG signal if the following prerequisites meet: (1) the sign of electrical activity produced by each neuron should be same; (2) the specific axis of electrical activity generated by most of the neurons should be perpendicular to the scalp; (3) neuronal synchrony of neurons should be high; (4) neuronal dendrites should be aligned in parallel to summate the potential which results into a production of signal and this signal could be detectable at some distance. Therefore finding patterns for neuronal communication is a complex task. Nevertheless, there exist some characteristics of EEG, which could be the basis of BCI system: (1) rhythmic brain activity; (2) Event-related potentials (ERP); (3) Event-related synchronization (ERS) and Event-related desynchronization (ERD) [1–3].

Brain Rhythms Brain is always working and depending upon the perception level, it shows different rhythmic activity. The rhythms are affected by thoughts and preparation of actions, for example eye blink can attenuate particular rhythm. The reality that sheer thoughts distress the rhythms can become the basis for the BCI system. Different brain rhythms can be identified in EEG with different range of frequencies [22]. They have given Greek letters delta, theta, alpha, beta, gamma, and mu (δ, θ, α, β, γ, and μ) to represent them. The order and meaning of letters is not logical.

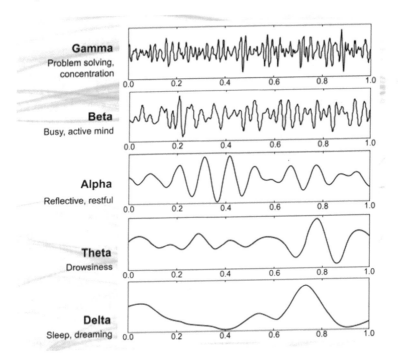

Fig. 5 Different brain waves [23]

Figure 5 is demonstrates different brain wave patterns available in brain electrical activity [23].

The delta wave can be recorded from 0.1 to 3.5 Hz of frequency range and with amplitude of 50–100 μV. This irregular rhythmic activity has found in infants (around 2 months) in waking stage. In adult's delta rhythm found only in deep sleep stage and below 3.5 Hz of frequency range. Hence this wave is not useful in BCI research. Next in the queue is theta wave whose frequency and amplitude ranges from 4 to 7.5 Hz and below 100 μV respectively [24]. It can be recording on the frontal midline area on scalp. It rarely found in children of age two or below in waking stage. In adults theta waves can be recorded in drowsiness and during the sleep especially in females. It can be blocked by the eye opening and disappear with the occurrence of alpha activity. It had been used in different applications like Quadcopter [25]. Alpha wave has already been used in many BCI applications. Its frequency ranges from 8 to 13 Hz and its amplitude varies but stays below 50 μV. It appears in EEG mostly over the posterior regions of the brain, mostly on the occipital areas. It can be seen clearly in EEG during the conditions of physical relaxation and relative mental inactivity. It can be attenuated by attention especially due to visual attention. Other important brain rhythm is mu rhythm. Its frequency and amplitude is same as alpha wave (10 Hz and below 50 μV respectively) but topographically and physiologically dissimilar from later one. This wave is present over the precentral motor cortex basically at

EEG C3, Cz and C4 electrode placement [7]. It can be blocked or attenuated when person perform motor activity or after training when person visualizing the motor activity. Instead of suppression, it shifts from ideal state to high frequency when motor action is performed. These facts about mu rhythm make it important in BCI research. Beta rhythm comes next in the list which ranges from 13 to 30 Hz and amplitude is around 30 μV. Beta is present over frontal and central region of brain. It is again divided into beta 1 (13–20 Hz), beta 2 (21–30 Hz) and gamma (30–60 Hz) [26]. Beta waves involve in conscious focus, problem solving, memorizing and tend to have a simulating effect. In adults it can be observed in awaken state while thinking and logical reasoning. It also plays an important role in BCI research. The summary of the brain rhythms are listed in Table 2.

Pineda [27] studied the use of the mu rhythm in BCI and concluded that "mu rhythm is not only modulated by the expression of self-generated movement but also by the observation and imagination of movement." Wolpaw and McFarland [28] have used the self-regulation of the mu rhythm or central beta rhythm amplitude in their BCI.

Event Related Potentials (ERP)

Event related potential recording technique is useful for human electrophysiology research. It has good and precise temporal resolution which can be the basis for testing the theories of perception, attention and cognition that are unobservable with behavioral methods. It allows recording the brain activity from 1 ms or above in the presence of stimuli or an event occurs. The potential changes are so small that in order to find the pattern, EEG samples are averaged. Further event-related potentials can be alienated into exogenous and endogenous depending upon the temporal resolution. It is exogenous potentials if resolution is under 100 ms and endogenous potentials occur after 100 ms onwards after the stimulus onset. They depend upon the properties of stimulus, physiological and behavioral processes related to the event. The main characteristics of ERP are polarity (positive or negative going signals), sensitivity to task manipulation, spatial distribution and time. Figure 6 is showing ERP generated in response to visual as well as audio stimuli presented to user [29].

P300 is most commonly explored ERP. This positive component of ERP occurs in brain at peak 300 ms or more (up to 900 ms) after onset stimuli. As it peaks above 100 ms, it is an endogenous ERP activity. A P300 based BCI system have advantage of minimal user training. In this system, users have to choose the one of the choices given in stimulus and designate this as the target. Evoked potentials (EP) are the subset of ERPs caused by the sensory stimulation in response of in physical stimulus (auditory, visual, somatosensory etc.). It ranges from 1 μV to few microvolts. They are present at different areas of brain like cerebral cortex, brain stem, spinal cord, peripheral nerves. Visual evoked potentials (VEP), auditory evoked potentials (AEP), steady state evoked potentials (SSEP) are some typical evoked potentials that reflects the output features of pathways of different brain sensory activities. Thought-translation

Table 2 Different brain rhythms [29]

Brain wave	Frequency range (in Hz)	Amplitude (in μV)	Brain area	Found in age group	Use in BCIs
Delta (δ)	0.1–3.5	50–100	Cerebral cortex	Infants (2 months) in waking state, deep sleep stage of adults	No
Theta (θ)	4–7.5	Below 100	Frontal midline region	In Infants wake up stage; In adults in drowsiness and sleep stage	Yes
Alpha (α)	8–13	Below 50	Occipital area	In children 3 years (8 Hz); In adults (10 Hz) During eyes closed and under relaxation and relative mental inactivity	Yes
Mu (μ)	Below 10 Hz	Below 50	Motor cortex and somatosensory cortex	All ages; thought of movement or in the presence of light tactile stimuli	Yes
Beta (β)	13–30	Above 30	Frontal and central regions	In adults while cognitive task related to stimulus assessment and decision making	Yes

Mu and alpha waves is topologically and physiologically different from each other

device (TTD), a training device and spelling program was developed by Birbaumer et al. [3], for completely paralyzed patients using slow cortical potentials.

Event-Related Desynchronization (ERD) and Event Related Synchronization (ERS)

Event related desynchronization is decrease in certain rhythms due to movement or preparation of movement. Contrary to this, increase in the amplitude of the rhythm results event-related. Mostly mu and beta rhythms are the rhythms involved for ERD

Fig. 6 Different evoked potentials present in brain electrical activity [29]

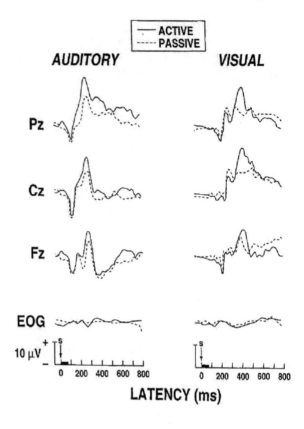

and ERS. ERD and ERS can be presented in both spatial and time domain. ERS/ERD can be measured by calculating the amplitude of certain brain wave before and after the presence of external/internal stimulus over a number of EEG trials. Then averaged power over a number of trials is measured in terms of percentage in relation to power of referential interval e.g. 1 s interval i.e. between 3.5 and 2.5 s before and after the event.

The interval between the two events should be random and not shorter than second to keep power at reference interval. In 1990, Pfurtscheller and Berghold [24] has developed Graz-BCI mu rhythm ERD/ERS based system using imagery of motor action as the mental task. Generalized ERS and ERS w.r.t. constant referencing scheme has been demonstrated in Fig. 7 [30].

Electrocorticogram (ECoG)

Electrocorticogram are the signals recorded at the surface of brain by placing the electrodes at the surface of cortex [31]. The surgical procedure "craniotomy" is used for opening the skull and cutting the membrane which covers the brain. ECoG

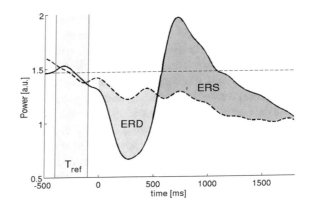

Fig. 7 Generalized ERS and ERS w.r.t. constant referencing scheme [30]

signals are like EEG signals but have better spatial resolution and attenuation due to absence of skull and scalp. The location and arrangement of electrodes as well as implant duration is variable and depends solely upon the application requirements. The electrodes used for recoding are typically platinum electrodes with 4 mm diameter and arranged in a grid of 8 × 8 or in strip of 4–6 electrodes. The distance between the electrodes is more often 10 mm. The spatial resolution and amplitude of ECoG signals vary from 1.25 to 1.4 mm and 50 to 100 μV respectively. Also, they are less affected by the brain and non-brain artifacts as the task related signals are larger than the noise floor of the amplifier/digitizer. Thus, the signal to noise ratio of ECoG signals is much higher than the EEG signals. It also concludes that they carry substantial amount of information about cognitive, motor and language tasks. The brain neurons stays undamaged as the electrodes do not penetrate the brain. From the literature [32], it can be concluded that ECoG electrodes are likely to provide longer stability than fully invasive intracortical electrodes. In spite of its advantages over EEG and intracortical recording, ECoG signals generally are not used for research need as there major surgery is involved. Typically used for medical implications especially for actual site and extent of epilepsy symptoms [33]. Perhaps, the future of nanotechnologies that might develop nano-detectors to be implanted inertly in the brain, may provide a definite solution to the problems of long-term invasive applications. Further, a link between the microelectrode and external hardware that uses wireless technology is needed to reduce the risks of infection. Wireless transmission of neuronal signals has already been tested in animals [34–36]. Further refinements of recording and analysis techniques will probably increase the performance of both invasive and non-invasive modalities.

3.1.2 Preprocessing of Acquired EEG Signals

Digital EEG data recordings have advantages of flexibility, user specific montage selection, horizontal scaling likes compression and time resolution, filters, vertical scaling of sensitivity etc. EEG data recordings are digital time series or set of discrete time series, thus it makes possible application of variety of digital signal processing techniques. Raw EEG data is contaminated with the other neurological or non-neurological signals which are known as artifacts e.g. eye blink, muscle activity, electrode movement etc. [37, 38]. Electromyogram (EMG) is class of artifacts due to muscle activity like facial movement, tongue movement, neck movement etc. the noise created by eye blink are electroculogram (EOG) signals and have high amplitude then neural signals. These artifacts results into interference in control signals for BCIs, poor signal to noise ratio (SNR) and change in the distinctiveness of specific interest of EEG data. Thus, removal of the artifacts from raw EEG data is necessary for improved SNR signals and BCI performance.

The EEG signal must be amplified, filtered, digitized and referenced before extracting the features out of it. There are many signal preprocessing methods exists, only EEG signal preprocessing methods are discussed here. The EEG signal must be boosted, amplified from few microvolt signals to million-fold to avoid the artifacts. The amplified signal then filtered in the range of 0.5–50 Hz to include necessary oscillatory components of EEG and to filter out high frequency signals like muscle activities (EMG) (>50 Hz), eye blinks (EOG) etc. Most of the researchers have used subject dependent band filter to filter the raw EEG signal like Notch, Finite Impulse response etc. [39–44]. This method is also known as temporal filtering and eliminates low as well as high and frequencies from signal. Signals can be spatially filtered using referencing schemes like common average referencing (CAR) [45], bipolar referencing, surface Laplacian. These filters use high pass spatial filtering to enhance the focal activity like mu and beta rhythms from local sources. The authors [46, 47] has used subject specific filtering using Independent Component Analysis (ICA) for blind source separation which assumes EEG data as linear superposition of independent components to remove the artifacts. The artifacts fNIRS due to breathing and heart beat can be filtered by moving average filters [48], IIR low pass filters [43], wavelet denoising [49].

3.1.3 Feature Extraction

The identification of signal's characteristics (features) that might help in identifying the specific pattern related to user intends present in filtered, amplified, digitized and referenced EEG signal is known as feature extraction process in BCI design. These features can be the basis of pattern recognition algorithms that leads to classification of mental activity [7]. The aim of feature extraction step is to find most distinctive features and thus, enhancing the signal to noise ratio (SNR). This important step becomes difficult when signals and noise are similar e.g. EOG is very similar to beta rhythms and EMG is very similar to slow cortical potentials (SCPs). EEG signals are spread over space, time and frequency. It can be studied in many domains like time domain, frequency domain or time-frequency domain. Bashashati et al. [50] reviewed different types of feature extraction methods in 2007. Many features like amplitude values of signal, auto regressive model coefficients (AR), band power, power spectrum density (PSD), correlation coefficients, entropy, wavelet coefficients etc. are studied and proven to be good for pattern matching algorithms. **Common spatial patterns (CSP)** are the most of efficient method for feature extraction from EEG signals [51]. Several variants of CSP method established for grasping spatial information of brain signals like Probabilistic common spatial patterns [52], bank regularized common spatial pattern ensemble [53]. Signal power/energy levels at different location over the scalp are known as **band power (BP)** features [1, 2]. After band power estimation of signals, these values can be used to find the event related synchronization/event related desychronization (ERS/ERD) maps to visualize certain activity/events in the signal. The raw signal should be band passed filtered within defined frequency bands and then squared and then averaged for consecutive time intervals. Visualize ERS/ERD for these values for each subject and then selection of bands with most distinctive information is stored for further classification [6]. The authors [54] has compared the CSP and BP features for four class BCI experiment and tackled the BP feature by adding phase information with time information. **Power Spectrum Density (PSD)** is the power distribution with frequency in signals/time series. The power of a signal can be power only or can be squared value of signal. The PSD feature only exists if the signal is wide-sense stationary process. PSD is the Fourier transform (FT) of autocorrelation of the wide-sense stationary signal. It does not exist in non-stationary signals as autocorrelation function must have two variables. However some researchers have estimated time varying spectral density as distinctive feature [55]. **Autoregressive (AR)** model coefficients also have shown good results for classification of different mental task/event using EEG signals [56–59]. The linear regression of current series data against one or more prior series of data is used to find autoregressive model coefficients. Many variants of linear regression can be applied for estimation of autoregressive like least square regression, recursive-least-square methods etc. Another Burg method is well known method for estimating reflection coefficients for autoregressive models. Differentials **Entropy** is also used as distinctive feature by authors of [60, 61]. Moreover **Wavelet Coefficients** also have been employed to extract features for EEG signal classifications [62–64]. The wavelet fuzzy approximate entropy, clustering techniques, cross-correlation techniques and many techniques exists for feature extraction from raw EEG signals. Following are some discussion points that might be of interest in deciding the feature to be used:

- **Usage of BCI**: BCI can be used as online or offline. Feature extraction for designing online BCI application is more complex than offline BCI design. Thus, low complexity features within small time frame would be advised choice for the design.
- **Robust BCI**: the noisy EEG signals have poor SNR and more sensitive to outliers. Thus, robustness towards artifacts and noise must be taken care for the BCI design.
- **Distinctiveness**: higher distinctiveness of extracted features towards brain events, easier and accurate is the classification task. This uniqueness can be measured with measure/index e.g. Fisher Index, DBI [6] or direct accuracy of classifier.
- **Non-stationarity**: for designing online BCI systems, non-stationarity based time varying shift detection in intra or inter session changes of EEG data could be a

point of interest. Some features like approximate entropy is less affected by these shift variation in EEG signals.

The choice of features and application of the BCI system are correlated. Feature can be ignored/selected on the basis of application of BCI system. Traditional feature extraction techniques like AR model, PSD or band power assumes the EEG signal as superimposition independent wave (mostly sinusoidal) components and avoid the phase information. Higher order statistics and non-linear feature extraction can be used to tackle this problem [65, 66].

3.1.4   Feature Selection

The features extracted can be high dimension vectors depends upon the number of channels, number of trials, number of sessions from multiple modality and sampling rate of modality. It is neither realistic nor useful to consider all features for classification. So selecting a smaller subset of distinctive feature set or feature space projection is an important step in pattern recognition for classification. The aim of feature selection process is to remove the redundant and uninformative features along with finding unique features which do not over fit the training set and classify the real dataset with higher accuracy even in the presence of noise and artifacts [67]. Projection techniques can be useful when the relevant information is spread in all over feature space and data is transformed in order to retrieve the discriminative information. In some applications channel selection might be helpful by setting the score to features of different channels. Then, channels having features with highest score is selected for further classification. Thus, there could be three approaches to handle the problem of high dimensionality:

- **Feature Selection**: here the goal is to find best combination of subset features using search base methods like genetic algorithms, wrapper's approach, filter approach, Sequential forward floating search etc. there is basic two criteria to find the good feature set (1) an optimized search method (2) a performance measure to evaluate the selected subset of features searched by (1). Finding the appropriate subset of features is considered as NP-hard [68]. Figure 8 depicts the four stage feature selection process demonstrated by authors Liu and Yu [69].

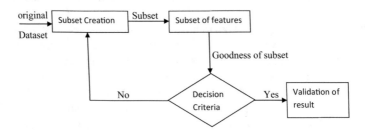

Fig. 8 Feature selection steps [69]

Both heuristic and intuitive search methods can be used for the searching purpose. Based on these factors, Wrapper approach and filter approach can be used to evaluate the performance of feature subset. In wrapper approach classifier is defined first, takes subset of feature as an input to classifier for training, then classification accuracy is evaluated in validation testing phase and finally these accuracies are compared across each subset. On the other hand, filter approach evaluate the goodness of features on the basis of measures/indexes independent of classifier. Distance measures in cluster like Davies-Bouldin Index (DBI) [6], information measures like information gain, dependency measure like coefficient of correlated feature or similarity index etc. There is a hybrid approach which uses both wrapper and filter to reach the higher accuracy in less computational cost [65].

- **Dimensionality Reduction**: here reduction of feature space is done by projecting high dimensional features into lower dimensional feature space. To deal with this curse of dimensionality, these methods can be divided into categories like linear/non-linear, supervised/unsupervised. Linear methods like principal component analysis (PCA), factor analysis (FA) consider covariance of data and transform it linearly to reduce the dimensions of observable random variables. Most nonlinear unsupervised method for dimensionality reduction is based on manifold learning theory. In these methods a weighted graph of data points depending upon the neighboring relation, are projected into lower dimensional space [65]. These methods uses structural knowledge like locality or proximity relation while maintain the relationship among the data points. These methods can be categorized in the following three methods [70]: (1) methods which preserve local properties of data in lower dimension e.g. Isomap, Kernal PCA (2) method which preserve global properties of data in lower dimension e.g. Laplacian Eigenmaps (3) methods which align mixture of linear models globally e.g. Manifold Charting.
- **Channel Selection**: here main aim is to find combination of channels which are generating most relevant and distinctive information specific to application. In some cases these methods are advantages than feature selection methods e.g. finding the spatial distribution of motor imagery events. The first approach for channel selection is to apply the feature selection methods and then mapping these features with associated channels. This method is limited to some specific applications. On the other hand, direct channel selection incorporate prior knowledge into analysis of results or in the selection process which leads to better understanding of spatial information, further can be used to implement required control.

3.1.5 Pattern Matching

The ultimate goal of BCI design is to translate the mental event of user into control commands. The acquired raw EEG signal has to be converted into real action in surrounding environment. So, classification or pattern matching of the signal into

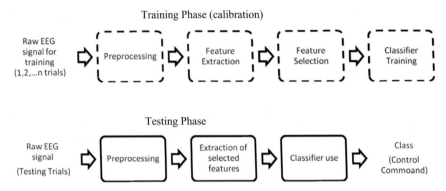

Fig. 9 Typical classification approach for EEG based BCI design

predefined classes is naturally the next step after preprocessing and feature extraction and selection. Machine learning has played an important role not only in identifying the user intent but also handle the variation in ongoing user's signals. Considering traditional approach of pattern matching [71], the classification algorithms for mental task recognition inside the EEG signals can be categorized in four categories: (1) adaptive classifiers, (2) transfer learning based classifiers, (3) matrix and tensor classifiers and (4) deep learning based classifiers.

- **Adaptive Classifiers**: In mid-2000s adaptive classifiers were used for EEG based BCI design [72–74]. The adaptive classifiers update the parameters (e.g. weights, error) incrementally over time and classifiers adapt the changes in the incoming EEG data. This enables the classifier work efficiently even if there is drift in the dataset. These classifiers can use supervised or unsupervised adaption [75, 76]. The former adaption uses previous knowledge of output classes. Figure 9 demonstrates the typical supervised classification approach for EEG-based BCI design. The dotted lines denote the algorithm which can be optimized from available data in training phase. The optimized algorithms then can be used for testing phase or original use to translate electrical brain signals into real time control commands. The real time or free BCI cannot take advantage from supervised adaption techniques as the true label of raw EEG data is unknown. Whereas, the unsupervised adaption approach do not use any previous knowledge of output classes and thus, output labels are unknown. The class label estimation can be done based on retraining/updating of classifiers or adaption with unknown class labels e.g. by updating mean or correlation matrix of variables. The combination of both type of adaption is known as semi-supervised adaption. These adaptions consider both the unlabeled and labeled dataset for training the classifier. First the classifier is trained with available dataset along with output class label. Then unlabeled testing data is classified by this supervised trained dataset. Finally, classifier is retrained/updated incrementally with unlabeled and available labeled dataset.

Various state-of-art classification algorithms have been employed by different groups to infer the mental task. Linear discriminant analysis (LDA), quadratic discriminant analysis (QDA) [72], adaptive Bayesian classifier [77], adaptive support vector machine (SVM) [78, 79], adaptive probabilistic neural network [80], radial basis function (RBF) kernels [81], L2-regularized linear logistic regression classifiers [82] are combination of linear or nonlinear state-of-art algorithms for supervised adaption approach. Ensemble and extreme leaning has also been implemented by Li and Zhang [83]. The unsupervised learning is complex and difficult to implement due unavailability of class specific information. Adaptive LDA and Gaussian Mixture model (GMM) [84], Adaptive LDA with Fuzzy C-means [85], Incremental logistic regression [86], Incremental SVM [87], Semi-supervised SVM [88], Unsupervised linear classifier [89] are some semi- or unsupervised algorithms used in different modalities in BCI design.

- **Matrix and Tensor Based Classifiers**: These classifiers works on the alternate approach as used for adaptive classifiers i.e. feature extraction and then selecting the relevant features. Instead of optimizing dual problem, these classifiers do the mapping of the data directly to classification domain e.g. geographical space. The idea behind these classifiers is the assumption that spatial distribution and power can be considered fixed and thus, can be represented in covariance matrices. These covariance matrices can be used directly as an input to classifier. Figure 10 demonstrate both adaptive feature learning and direct learning of matrices approaches for pattern matching in EEG signal classification. This approach can be applied to

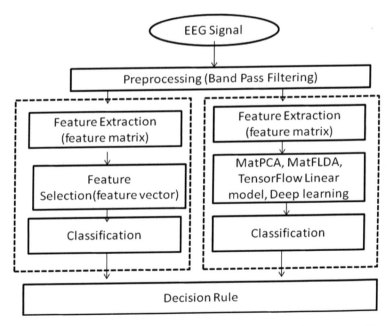

Fig. 10 Two approaches for classification of EEG data. The dotted area is interchangeable

both oscillation based BCI and ERP based BCI systems. A regularized discriminative framework for EEG analysis in which data is represented as augmented covariance matrices has used this approach [90]. Riemannian Geometry Classifiers (RGC) in [91] are also based on the same concept in which data is directly mapped into geographical space with suitable matrices. These approaches benefits in the form of higher accuracy but complexity, high dimensionality of these classifiers is more demanding than tradition approaches.

Tensors are multi-way arrays and used to generate high order tensors from EEG data format. For example, 3rd-order tensor for EEG classification can be represented as space × frequency × time. These modes define the order of tensor, also known as dimensions of tensors. Almost all classification algorithms can be generalized using tensors but this field is yet be explored [92, 93].

- **Transfer Learning**: The hypothesis which most of the machine learning algorithms follows that the data set for training and testing belongs to same data domain with same probability domain. Opposite to this hypothesis, in BCI design data distribution is different in real time testing phase across time or subject. Transfer learning handles this problem by exploiting the knowledge about one task, while learning another related task. So, effectiveness of transfer learning is totally depends upon the correlation in these task. For instance, motor imagery task performed by two subjects is more effective than performing motor imagery task and p300 speller task by same subject. Transfer learning plays important role where domain data is labeled for one task, and target domain contains the scarce to acquire another task. Transfer learning can be categories in two types based upon domains, tasks and learning setting. **Homogeneous transfer learning** is the learning where source domain task and target domain task is same, and adaption of the probability distribution or conditional probability distribution is not same in source and target domain. Whereas, **Inductive transfer learning** is where source task and target task are different in labeled data in both source and target domain. For instance, there could be left hand and right hand movement is labeled in both source and target domain, whilst target domain involves tongue movement. Another situation, **Transductive transfer learning** is the situation where source and target domain are different but tasks are similar. It happens frequently in BCI systems, as there is inter/intra session variability or inter-subject variability usually arises.

Many transfer learning approaches evolve by transformation of data to match their distribution. This could be linear or non-linear transformation. Figure 11 illustrates an example of domain adaption and transfer learning where source domain and target domain are differently labeled. A normal classifier trained on source domain will perform poorly on target domain. But by applying domain adaption technique [94] transfer the dataset distribution as to match the source and target domain distribution. A detailed survey has been presented by Pan and Yang [95] on transfer learning for more detailed illustration on transfer learning.

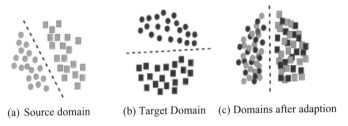

(a) Source domain   (b) Target Domain   (c) Domains after adaption

Fig. 11 Domain adaption [94] in transfer learning

- **Deep Learning**: is special branch of machine learning algorithms which directly learn from the data set instead of learning from extracted feature set. It is based on the deep learning done by the human brain which created the pattern from data and learn from it for decision making. In recent year deep learning has shown good classification results and improved accuracy of the pattern recognition system. Like machine learning, it is also supervised, unsupervised or semi-supervised. An inbuilt cascade of feature extractor modules handles the non-linearity of available data domain. Figure 12 demonstrates the difference between tradition machine learning algorithms and deep learning algorithms.

Deep Boltzmann Machine (DBM), Recurrent Neural Network (RNN), Recursive Neural Network (RvNN), Deep Belief Network (DBN), Convolution Neural Network (CNN), and Auto Encoder (AE) are some examples of deep learning algorithms.

Deep Extreme Learning Machine (ELM) has used by authors of [96] for finding slow cortical potentials (SCP) in EEG signals. This ELM contains multilayer of extreme learning machine ending with last layer of kernel ELM. The motion onset visual evoked potential BCI features have been extracted using deep brief network (DBN). The DBN deep learning machine is composed of three Restricted Boltzmann machine (RBM) [97]. Yin and Zhang [98] employed adaptive deep neural network (DNN) to classify both workload as well as emotions. They compose the stack of Auto Encoder (AE). They retrained the first layer of network with adaptive learning algorithm taking labeled input with estimated class.

The deep learning classifiers are advantageous as it leads to better features and classifying accuracy. But they need large number of training dataset for calibration.

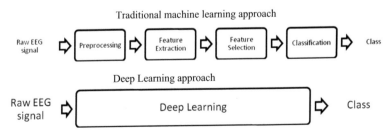

Fig. 12 Traditional versus deep learning approach

BCI is user specific application, subject have to perform thousands of relevant task for calibration before actual use of it. For online systems, it is quite expensive in terms of money as well as time.

3.1.6   System Feedback and User Training

Finally, before providing feedback to user about a specific mental state is recognized or not, EEG signals should be classified on the basis of selected features to convert the EEG signal into a control command. Thus, system feedback and user training is an important step in BCI design. Many research findings have shown that inaccurate feedback to user causes the impeded accuracy of BCI system [18]. Feedback can be continuous/discrete audio video signal, virtual/realistic 1D, 2D and 3D environment. Feedback makes the BCI design as adaptive closed loop system between human brain and computer.

4   BCI Performance Measures

Evaluation of BCI system is different depending upon the design of BCI system and target application. Some of the common BCI performance measures are classification accuracy, kappa metric, bit rate, area under the curve (AUC), uncertainty and mutual information, the receiver operating characteristic (ROC) curve and entropy. Every step of BCI design has different components for performance evaluation in closed loop BCI dependent upon the design. The basic and most commonly used method is classification accuracy specifically for the equally distributed samples per class and for unbiased classifiers [77–81]. Another Kappa metrics or the confusion matrix is used to measure the sensitivity-specificity pair for unbalanced classes and less biased data [54]. A bit rate, an information transfer rate is used in account to both accuracy and speed of a BCI [99]. Channel capacity has to be calculated with several assumptions in bits/min. Entropy and uncertainty of a classifier can also be used to appraise the performance of a BCI system [60].

5   Conclusion

This new pathway to human brain can open many doors to complex and unimaginable solutions to many applications. Many more type of diseases can be diagnosed. There is a great scope of enhancement in existing BCIs using artificial intelligence and machine learning algorithms. Use of high computing electronic devices and transfer learning, tensor and deep learning algorithms could serve the purpose. Security and privacy issues open challenge has gained significant attention and can further be explored [100].

References

1. N. Birbaumer, N. Ghanayim, T. Hinterberger, I. Iversen, B. Kotchoubey, A. Kübler, J. Perelmouter, E. Taub, H. Flor, A spelling device for the paralysed. Nature **398**(6725), 297 (1999)
2. G. Pfurtscheller, D. Flotzinger, J. Kalcher, Brain-computer interface—a new communication device for handicapped persons. J. Microcomput. Appl. **16**(3), 293–299 (1993)
3. N. Birbaumer, T. Hinterberger, A. Kubler, N. Neumann, The thought-translation device (TTD): neurobehavioral mechanisms and clinical outcome. IEEE Trans. Neural Syst. Rehabil. Eng. **11**(2), 120–123 (2003)
4. S.G. Mason, G.E. Birch, A general framework for brain-computer interface design. IEEE Trans. Neural Syst. Rehabil. Eng. **11**(1), 70–85 (2003)
5. J.R. Wolpaw, N. Birbaumer, W.J. Heetderks, D.J. McFarland, P.H. Peckham, G. Schalk, E. Donchin, L.A. Quatrano, C.J. Robinson, T.M. Vaughan, Brain-computer interface technology: a review of the first international meeting. IEEE Trans. Rehabil. Eng. **8**(2), 164–173 (2000)
6. B. Blankertz, G. Dornhege, M. Krauledat, K.R. Müller, G. Curio, The non-invasive Berlin brain–computer interface: fast acquisition of effective performance in untrained subjects. NeuroImage **37**(2), 539–550 (2007)
7. J.R. Wolpaw, N. Birbaumer, D.J. McFarland, G. Pfurtscheller, T.M. Vaughan, Brain–computer interfaces for communication and control. Clin. Neurophysiol. **113**(6), 767–791 (2002)
8. A. Kübler, B. Kotchoubey, J. Kaiser, J.R. Wolpaw, N. Birbaumer, Brain–computer communication: unlocking the locked in. Psychol. Bull. **127**(3), 358 (2001)
9. J. del R. Millan, J. Mouriño, M. Franzé, F. Cincotti, M. Varsta, J. Heikkonen, F. Babiloni, A local neural classifier for the recognition of EEG patterns associated to mental tasks. IEEE Trans. Neural Netw. **13**(3), 678–686 (2002)
10. A. Kostov, M. Polak, Parallel man-machine training in development of EEG-based cursor control. IEEE Trans. Rehabil. Eng. **8**(2), 203–205 (2000)
11. S.J. Roberts, W.D. Penny, Real-time brain-computer interfacing: a preliminary study using Bayesian learning. Med. Biol. Eng. Compu. **38**(1), 56–61 (2000)
12. C.W. Anderson, Z. Sijercic, Classification of EEG signals from four subjects during five mental tasks, in *Solving Engineering Problems with Neural Networks: Proceedings of the Conference on Engineering Applications in Neural Networks (EANN'96)*, June 1996 (Turkey), pp. 407–414
13. J. Kaiser, J. Perelmouter, I.H. Iversen, N. Neumann, N. Ghanayim, T. Hinterberger, A. Kübler, B. Kotchoubey, N. Birbaumer, Self-initiation of EEG-based communication in paralyzed patients. Clin. Neurophysiol. **112**(3), 551–554 (2001)
14. Z. Tang, C. Li, S. Sun, Single-trial EEG classification of motor imagery using deep convolutional neural networks. Optik-Int. J. Light Electron Opt. **130**, 11–18 (2017)
15. M. Pavlova, W. Lutzenberger, A. Sokolov, N. Birbaumer, Dissociable cortical processing of recognizable and non-recognizable biological movement: analysing gamma MEG activity. Cereb. Cortex **14**(2), 181–188 (2004)
16. H.J. Chizeck, T. Bonaci, Brain-computer interface anonymizer. U.S. Patent Application 14/174,818, University of Washington, 2014
17. B. Blankertz, C. Sannelli, S. Halder, E.M. Hammer, A. Kübler, K.R. Müller, G. Curio, T. Dickhaus, Neurophysiological predictor of SMR-based BCI performance. NeuroImage **51**(4), 1303–1309 (2010)
18. Á. Barbero, M. Grosse-Wentrup, Biased feedback in brain-computer interfaces. J. Neuroeng. Rehabil. **7**(1), 34 (2010)
19. M. Tangermann, K.R. Müller, A. Aertsen, N. Birbaumer, C. Braun, C. Brunner, R. Leeb, C. Mehring, K.J. Miller, G. Mueller-Putz, G. Nolte, Review of the BCI competition IV. Front. Neurosci. **6**, 55 (2012)
20. R. Oostenveld, P. Praamstra, The five percent electrode system for high-resolution EEG and ERP measurements. Clin. Neurophysiol. **112**(4), 713–719 (2001)

21. B. Allison, *Brain Computer Interface Systems* (1999). http://bci.ucsd.edu/
22. E. Niedermeyer, The normal EEG of the waking adult, in *Electroencephalography: Basic Principles, Clinical Applications, and Related Fields*, vol. 167, pp. 155–164 (2005)
23. P.A. Abhang, B.W. Gawali, S.C. Mehrotra, *Introduction to EEG- and Speech-Based Emotion Recognition* (Academic Press, 2016)
24. G. Pfurtscheller, A. Berghold, Patterns of cortical activation during planning of voluntary movement. Electroencephalogr. Clin. Neurophysiol. **72**(3), 250–258 (1989)
25. S. Rosca, M. Leba, A. Ionica, O. Gamulescu, Quadcopter control using a BCI. IOP Conf. Ser. Mater. Sci. Eng. **294**(1), 012048 (2018)
26. H. Liu, J. Wang, C. Zheng, P. He, Study on the effect of different frequency bands of EEG signals on mental tasks classification, in *2005 IEEE Engineering in Medicine and Biology 27th Annual Conference*, Jan 2006 (IEEE), pp. 5369–5372
27. J.A. Pineda, The functional significance of mu rhythms: translating "seeing" and "hearing" into "doing". Brain Res. Rev. **50**(1), 57–68 (2005)
28. J.R. Wolpaw, D.J. McFarland, Control of a two-dimensional movement signal by a noninvasive brain-computer interface in humans. Proc. Natl. Acad. Sci. **101**(51), 17849–17854 (2004)
29. J.Y. Bennington, J. Polich, Comparison of P300 from passive and active tasks for auditory and visual stimuli. Int. J. Psychophysiol. **34**(2), 171–177 (1999)
30. S. Lemm, K.R. Müller, G. Curio, A generalized framework for quantifying the dynamics of EEG event-related desynchronization. PLoS Comput. Biol. **5**(8), e1000453 (2009)
31. J.A. Wilson, E.A. Felton, P.C. Garell, G. Schalk, J.C. Williams, ECoG factors underlying multimodal control of a brain-computer interface. IEEE Trans. Neural Syst. Rehabil. Eng. **14**(2), 246–250 (2006)
32. T.G. Yuen, W.F. Agnew, L.A. Bullara, Tissue response to potential neuroprosthetic materials implanted subdurally. Biomaterials **8**(2), 138–141 (1987)
33. U.R. Acharya, F. Molinari, S.V. Sree, S. Chattopadhyay, K.H. Ng, J.S. Suri, Automated diagnosis of epileptic EEG using entropies. Biomed. Signal Process. Control **7**(4), 401–408 (2012)
34. T.A. Szuts, V. Fadeyev, S. Kachiguine, A. Sher, M.V. Grivich, M. Agrochão, P. Hottowy, W. Dabrowski, E.V. Lubenov, A.G. Siapas, N. Uchida, A wireless multi-channel neural amplifier for freely moving animals. Nat. Neurosci. **14**(2), 263 (2011)
35. M. Berger, A. Gail, The Reach Cage environment for wireless neural recordings during structured goal-directed behavior of unrestrained monkeys. bioRxiv 305334 (2018)
36. V. Cutsuridis, Memory prosthesis: is it time for a deep neuromimetic computing approach? Front. Neurosci. **13** (2019)
37. M. Fatourechi, A. Bashashati, R.K. Ward, G.E. Birch, EMG and EOG artifacts in brain computer interface systems: a survey. Clin. Neurophysiol. **118**(3), 480–494 (2007)
38. Á. Costa, R. Salazar-Varas, A. Úbeda, J.M. Azorín, Characterization of artifacts produced by gel displacement on non-invasive brain-machine interfaces during ambulation. Front. Neurosci. **10**, 60 (2016)
39. H. Wang, Y. Li, J. Long, T. Yu, Z. Gu, An asynchronous wheelchair control by hybrid EEG–EOG brain–computer interface. Cogn. Neurodyn. **8**(5), 399–409 (2014)
40. J. Jiang, Z. Zhou, E. Yin, Y. Yu, D. Hu, Hybrid brain-computer interface (BCI) based on the EEG and EOG signals. Bio-Med. Mater. Eng. **24**(6), 2919–2925 (2014)
41. S.R. Soekadar, M. Witkowski, N. Vitiello, N. Birbaumer, An EEG/EOG-based hybrid brain-neural computer interaction (BNCI) system to control an exoskeleton for the paralyzed hand. Biomed. Eng./Biomed. Tech. **60**(3), 199–205 (2015)
42. S.R. Soekadar, M. Witkowski, C. Gómez, E. Opisso, J. Medina, M. Cortese, M. Cempini, M.C. Carrozza, L.G. Cohen, N. Birbaumer, N. Vitiello, Hybrid EEG/EOG-based brain/neural hand exoskeleton restores fully independent daily living activities after quadriplegia. Sci. Robot. **1**(1), eaag3296–1 (2016)
43. W.L. Zheng, B.L. Lu, A multimodal approach to estimating vigilance using EEG and forehead EOG. J. Neural Eng. **14**(2), 026017 (2017)

44. M.H. Lee, J. Williamson, D.O. Won, S. Fazli, S.W. Lee, A high performance spelling system based on EEG-EOG signals with visual feedback. IEEE Trans. Neural Syst. Rehabil. Eng. **26**(7), 1443–1459 (2018)
45. B. Blankertz, R. Tomioka, S. Lemm, M. Kawanabe, K.R. Muller, Optimizing spatial filters for robust EEG single-trial analysis. IEEE Signal Process. Mag. **25**(1), 41–56 (2007)
46. J. Shin, K.R. Müller, C.H. Schmitz, D.W. Kim, H.J. Hwang, Evaluation of a compact hybrid brain-computer interface system. BioMed Res. Int. (2017)
47. B.J. Culpepper, R.M. Keller, Enabling computer decisions based on EEG input. IEEE Trans. Neural Syst. Rehabil. Eng. **11**(4), 354–360 (2003)
48. B. Koo, H.G. Lee, Y. Nam, H. Kang, C.S. Koh, H.C. Shin, S. Choi, A hybrid NIRS-EEG system for self-paced brain computer interface with online motor imagery. J. Neurosci. Methods **244**, 26–32 (2015)
49. M.J. Khan, M.J. Hong, K.S. Hong, Decoding of four movement directions using hybrid NIRS-EEG brain-computer interface. Front. Hum. Neurosci. **8**, 244 (2014)
50. A. Bashashati, M. Fatourechi, R.K. Ward, G.E. Birch, A survey of signal processing algorithms in brain–computer interfaces based on electrical brain signals. J. Neural Eng. **4**(2), R32 (2007)
51. K.P. Thomas, C. Guan, C.T. Lau, A.P. Vinod, K.K. Ang, A new discriminative common spatial pattern method for motor imagery brain–computer interfaces. IEEE Trans. Biomed. Eng. **56**(11), 2730–2733 (2009)
52. W. Wu, Z. Chen, X. Gao, Y. Li, E.N. Brown, S. Gao, Probabilistic common spatial patterns for multichannel EEG analysis. IEEE Trans. Pattern Anal. Mach. Intell. **37**(3), 639–653 (2014)
53. S.H. Park, D. Lee, S.G. Lee, Filter bank regularized common spatial pattern ensemble for small sample motor imagery classification. IEEE Trans. Neural Syst. Rehabil. Eng. **26**(2), 498–505 (2017)
54. G. Townsend, B. Graimann, G. Pfurtscheller, A comparison of common spatial patterns with complex band power features in a four-class BCI experiment. IEEE Trans. Biomed. Eng. **53**(4), 642–651 (2006)
55. M.P. Norton, D.G. Karczub, *Fundamentals of Noise and Vibration Analysis for Engineers* (Cambridge University Press, 2003)
56. A. Subasi, E. Erçelebi, A. Alkan, E. Koklukaya, Comparison of subspace-based methods with AR parametric methods in epileptic seizure detection. Comput. Biol. Med. **36**(2), 195–208 (2006)
57. A. Schlögl, G. Pfurtscheller, Considerations on adaptive autoregressive modelling in EEG analysis, in *Proceedings of First International Symposium on Communication Systems and Digital Signal Processing* (1998)
58. Y. Zhang, S. Zhang, X. Ji, EEG-based classification of emotions using empirical mode decomposition and autoregressive model. Multimed. Tools Appl. **77**(20), 26697–26710 (2018)
59. J.Y. Chang, M. Fecchio, A. Pigorini, M. Massimini, G. Tononi, B.D. Van Veen, Assessing recurrent interactions in cortical networks: modeling EEG response to transcranial magnetic stimulation. J. Neurosci. Methods **312**, 93–104 (2019)
60. D.W. Chen, R. Miao, W.Q. Yang, Y. Liang, H.H. Chen, L. Huang, C.J. Deng, N. Han, A feature extraction method based on differential entropy and linear discriminant analysis for emotion recognition. Sensors **19**(7), 1631 (2019)
61. R. Martín-Clemente, J. Olias, D. Thiyam, A. Cichocki, S. Cruces, Information theoretic approaches for motor-imagery BCI systems: review and experimental comparison. Entropy **20**(1), 7 (2018)
62. R. Elmahdi, N.Y. Amed, M.B.M. Amin, A.O. Hamza, S.A. Babaker, W.A.A. Elgylani, Comparative study between daubechies and coiflets wavelet decomposition mother families in feature extraction of BCI based on multiclass motor imagery discrimination. J. Clin. Eng. **44**(1), 41–46 (2019)
63. J. Zhou, M. Meng, Y. Gao, Y. Ma, Q. Zhang, Classification of motor imagery EEG using wavelet envelope analysis and LSTM networks, in *2018 Chinese Control and Decision Conference (CCDC)*, June 2018 (IEEE), pp. 5600–5605

64. Y. Wang, X. Li, H. Li, C. Shao, L. Ying, S. Wu, Feature extraction of motor imagery electroencephalography based on time-frequency-space domains. J. Biomed. Eng. **31**(5), 955–961 (2014)
65. M. Dyson, T. Balli, J.Q. Gan, F. Sepulveda, R. Palaniappan. 2008. Approximate entropy for EEG-based movement detection, pp. 110–115
66. S.M. Zhou, J.Q. Gan, F. Sepulveda, Classifying mental tasks based on features of higher-order statistics from EEG signals in brain–computer interface. Inf. Sci. **178**(6), 1629–1640 (2008)
67. K.R. Müller, M. Krauledat, G. Dornhege, G. Curio, B. Blankertz, Machine learning techniques for brain-computer interfaces. Biomed. Tech. **49**(1), 11–22 (2004)
68. G.H. John, R. Kohavi, K. Pfleger, Irrelevant features and the subset selection problem, in *Machine Learning Proceedings 1994* (Morgan Kaufmann, 1994), pp. 121–129
69. H. Liu, L. Yu, Toward integrating feature selection algorithms for classification and clustering. IEEE Trans. Knowl. Data Eng. **4**, 491–502 (2005)
70. L. Van Der Maaten, E. Postma, J. Van den Herik, Dimensionality reduction: a comparative. J. Mach. Learn. Res. **10**(66–71), 13 (2009)
71. P. Chaudhary, R. Agrawal, A comparative study of linear and non-linear classifiers in sensory motor imagery based brain computer interface. J. Comput. Theor. Nanosci. **16**(12), 5134–5139 (2019)
72. A. Schlögl, C. Vidaurre, K.R. Müller, Adaptive methods in BCI research—an introductory tutorial, in *Brain-Computer Interfaces* (Springer, Berlin, Heidelberg, 2009), pp. 331–355
73. P. Sykacek, S.J. Roberts, M. Stokes, Adaptive BCI based on variational Bayesian Kalman filtering: an empirical evaluation. IEEE Trans. Biomed. Eng. **51**(5), 719–727 (2004)
74. P. Shenoy, M. Krauledat, B. Blankertz, R.P. Rao, K.R. Müller, Towards adaptive classification for BCI. J. Neural Eng. **3**(1), R13 (2006)
75. M. Fernández-Delgado, E. Cernadas, S. Barro, D. Amorim, Do we need hundreds of classifiers to solve real world classification problems? J. Mach. Learn. Res. **15**(1), 3133–3181 (2014)
76. S. Marcel, J.D.R. Millán, Person authentication using brainwaves (EEG) and maximum a posteriori model adaptation. IEEE Trans. Pattern Anal. Mach. Intell. **29**(4), 743–752 (2007)
77. J.W. Yoon, S.J. Roberts, M. Dyson, J.Q. Gan, Adaptive classification for brain computer interface systems using sequential Monte Carlo sampling. Neural Netw. **22**(9), 1286–1294 (2009)
78. Q. Zheng, F. Zhu, J. Qin, P.A. Heng, Multiclass support matrix machine for single trial EEG classification. Neurocomputing **275**, 869–880 (2018)
79. A. Subasi, J. Kevric, M.A. Canbaz, Epileptic seizure detection using hybrid machine learning methods. Neural Comput. Appl. **31**(1), 317–325 (2019)
80. M.K. Hazrati, A. Erfanian, An online EEG-based brain–computer interface for controlling hand grasp using an adaptive probabilistic neural network. Med. Eng. Phys. **32**(7), 730–739 (2010)
81. T. Kawase, T. Sakurada, Y. Koike, K. Kansaku, A hybrid BMI-based exoskeleton for paresis: EMG control for assisting arm movements. J. Neural Eng. **14**(1), 016015 (2017)
82. X. Zhang, B. Hu, X. Ma, L. Xu, Resting-state whole-brain functional connectivity networks for MCI classification using L2-regularized logistic regression. IEEE Trans. Nanobiosci. **14**(2), 237–247 (2015)
83. J. Li, L. Zhang, Bilateral adaptation and neurofeedback for brain computer interface system. J. Neurosci. Methods **193**(2), 373–379 (2010)
84. B.A.S. Hasan, J.Q. Gan, Hangman BCI: an unsupervised adaptive self-paced brain-computer interface for playing games. Comput. Biol. Med. **42**(5), 598–606 (2012)
85. G. Liu, D. Zhang, J. Meng, G. Huang, X. Zhu, Unsupervised adaptation of electroencephalogram signal processing based on fuzzy C-means algorithm. Int. J. Adapt. Control Signal Process. **26**(6), 482–495 (2012)
86. A. Llera, M.A. van Gerven, V. Gómez, O. Jensen, H.J. Kappen, On the use of interaction error potentials for adaptive brain computer interfaces. Neural Netw. **24**(10), 1120–1127 (2011)
87. X. Artusi, I.K. Niazi, M.F. Lucas, D. Farina, Performance of a simulated adaptive BCI based on experimental classification of movement-related and error potentials. IEEE J. Emerg. Sel. Topics Circ. Syst. **1**(4), 480–488 (2011)

88. S. Lu, C. Guan, H. Zhang, Unsupervised brain computer interface based on inter-subject information, in *2008 30th Annual International Conference of the IEEE Engineering in Medicine and Biology Society*, Aug 2008 (IEEE), pp. 638–641
89. T. Verhoeven, D. Hübner, M. Tangermann, K.R. Müller, J. Dambre, P.J. Kindermans, Improving zero-training brain-computer interfaces by mixing model estimators. J. Neural Eng. **14**(3), 036021 (2017)
90. R. Tomioka, K.R. Müller, A regularized discriminative framework for EEG analysis with application to brain–computer interface. NeuroImage **49**(1), 415–432 (2010)
91. F. Yger, M. Berar, F. Lotte, Riemannian approaches in brain-computer interfaces: a review. IEEE Trans. Neural Syst. Rehabil. Eng. **25**(10), 1753–1762 (2016)
92. A.H. Phan, A. Cichocki, Tensor decompositions for feature extraction and classification of high dimensional datasets. Nonlinear Theory Appl IEICE **1**(1), 37–68 (2010)
93. Y. Washizawa, H. Higashi, T. Rutkowski, T. Tanaka, A. Cichocki, Tensor based simultaneous feature extraction and sample weighting for EEG classification, in *International Conference on Neural Information Processing*, Nov 2010 (Springer, Berlin, Heidelberg, 2010), pp. 26–33
94. S. Ben-David, T. Lu, T. Luu, D. Pál, Impossibility theorems for domain adaptation, in *International Conference on Artificial Intelligence and Statistics* (2010), pp. 129–136
95. S.J. Pan, Q. Yang, A survey on transfer learning. IEEE Trans. Knowl. Discov. Data Eng **22**(10) (2010)
96. S. Ding, N. Zhang, X. Xu, L. Guo, J. Zhang, Deep extreme learning machine and its application in EEG classification. Math. Probl. Eng. (2015)
97. T. Ma, H. Li, H. Yang, X. Lv, P. Li, T. Liu, D. Yao, P. Xu, The extraction of motion-onset VEP BCI features based on deep learning and compressed sensing. J. Neurosci. Methods **275**, 80–92 (2017)
98. Z. Yin, J. Zhang, Cross-session classification of mental workload levels using EEG and an adaptive deep learning model. Biomed. Signal Process. Control **33**, 30–47 (2017)
99. J. Kronegg, S. Voloshynovskyy, T. Pun, Analysis of bit-rate definitions for brain-computer interfaces (2005)
100. P. Chaudhary, R. Agrawal, Emerging threats to security and privacy in brain computer interface. Int. J. Adv. Stud. Sci. Res. **3**(12) (2018)

Printed in the United States
by Baker & Taylor Publisher Services